Hungry
Hollow

Hungry Hollow

The Story
of a Natural Place

A.K. DEWDNEY

COPERNICUS
AN IMPRINT OF SPRINGER-VERLAG

Published in the United States by Copernicus,
an imprint of Springer-Verlag New York, Inc.

Copernicus
Springer-Verlag New York, Inc.
175 Fifth Avenue
New York, NY 10010

Library of Congress Cataloging-in-Publication Data

Dewdney, A. K.
    Hungry Hollow : the story of a natural
place / A.K. Dewdney.
        p.   cm.
    ISBN 0-387-98415-1 (hardcover : alk. paper)
    1. Natural history–East (U.S.)   I. Title.
QH104.5.E37D48   1998
508.74–dc21   97-48857
                                                    CIP

Manufactured in the United States of America.
Printed on acid-free paper.
Designed by Karen Phillips.
Illustrations on frontispiece and on pages 3, 13, 31, 42, 70, 106, 112, 147,
and 184 by Christie Lyons. Illustrations on pages 127, 155, 162, 170,
205, and 216 by Roman Szolkowski.

9 8 7 6 5 4 3 2 1

ISBN 0-387-98415-1   SPIN 10659364

*This book is dedicated with affection and respect
to my biological mentors,
early and late, especially to*

Carl Robinow

Christopher Hannay

James Phipps

Contents

PREFACE   *ix*

Procyon lotor   *2*

The Tippecanoe Sea   *10*

Dianne sapiens   *17*

The Hackberry   *23*

The Ant's Journey   *30*

Congress of Birds   *40*

Microperson   *49*

Water   *61*

Cymbella and the Hypotrich   *68*

On the Back of a Turtle   *77*

The Meadow   *84*

The Labyrinths   *91*

Prayer of the Mantis   *98*

The Hydraulic Plant   *104*

The Storm   *110*

Abundance   *117*

In the Forest   *126*

The Art of Decay   *134*

Bear   *143*

Stories in Stone   *153*

Hungry Creek   *160*

Requiem for a Toad   *169*

Survival of the Lucky   *176*

The Book of Kaolinite   *183*

Didelphis virginianus   *191*

Animal Minds   *197*

Permanent Clearcut   *203*

Ursa Major   *210*

NOTES   *219*

BIBLIOGRAPHY   *230*

Preface

Hungry Hollow is nowhere and everywhere. An *identical* combination of geography and species will be found nowhere in the eastern deciduous forest region of North America, although many forest/floodplain environments within this region resemble it closely. In its generic form, to say nothing of its symbolic presence, Hungry Hollow will be found everywhere, standing for all natural places that have borne the brunt of human impact, whether from pollution, physical disturbance, or loss of habitat.

In its specific form, Hungry Hollow is an easy patchwork of three natural places near my home. From one of these it derives the river-valley setting, fossiliferous shale, and its name: The Hungry Hollow formation crawls with fossils, traces of an ancient mid-Devonian ecosystem. From a second place our Hungry Hollow gets its upland forest, and from a third its floodplain meadow.

In any attempt to describe this place, scientific terminology is not only inevitable but also desirable. Many scientific terms have a unique descriptive power—so much so that sometimes they are not even defined here! We also use the metric system, which has been universally adopted as the language of scientific measurement. Metric takes very little trouble to learn, but readers not used to this system may take the width of a fingernail as a centimeter and the height of their hips as a meter. A kilometer is about 60 percent of a mile.

Despite its limited size, a half-kilometer stretch of river valley, Hungry Hollow is a massive, complex ecosystem with a near-continuum of components. In a world where everything connects with everything else, it is impossible to linearize the description, by wedging it into a sequence of chapters, without overlooking the great majority of the connections. Within chapters, it demanded a fierce efficiency to portray the connections and interactions of Hungry Hollow. *Cymbella* is a diatom with her own story. But she has chlorophyll as well and stands for other algae and all plants. To do the place real justice, this book should have a thousand chapters, and every chapter should be a book!

The complexity of Hungry Hollow turns on its multidimensional structure. It consists not of one world, but of ten arranged like nesting boxes. Each world has one-tenth the dimension of its predecessor and yet overlaps it. Within World Zero, the world of ordinary human experience, things have dimensions of meters, more or less. From here we may readily appreciate World One, with its plants measured in decimeters, and World Two, with its birds measured in centimeters. But we will suffer eyestrain examining anything from World Three, where millimeters measure insects and mosses. World Four, ruled by tenths of a millimeter, is practically invisible to us. We simply do not see its nematodes and mites.

The worlds go on. World Five, which has the scale of tens of microns, is inhabited by *Paramecium* and *Spirogyra.* In World Six, microns measure the flagellate *Bodo,* the cyanobacterium called *Oscillatoria,* and the bacterium known as *Aeromonas.* Only the viruses—strange, partially living entities—are found in World Seven. At the nanometer scale of World Eight we find the structure of DNA, and Worlds Nine and Ten are made up of compounds and individual atoms, respectively. This is the first book about natural environments to mention these worlds specifically and to give them their due as the settings for natural places. It presents living components of all sizes, like so many chips of stained glass. It offers the reader a view of (or through) this magnificent natural window.

To those whose perspective exhibits what some biologists call the "macrofaunal bias," it may seem that microenvironments have been overdone in this book. Nothing of the kind. Despite the stress that this book places on them, they are still underemphasized!

Examining Hungry Hollow, with its thousands of species, also requires that we discuss taxonomy, and here again we must speak in scientific language. Although we all shrink to a certain degree from scientific names, there are excellent reasons to use them—and even to play with them a little.

Reason One: How better to recognize the finiteness of life on this planet than by realizing that there are only five main kingdoms (but see note on page 220) and, within them, no more than 130 phyla of living things? For each of the remaining taxonomic levels, from class down to species, the numbers of groups increase, but they still remain all too finite. In the end there are the individuals of each species, at any moment a definite number. As the numbers of individuals decline, so inevitably do the numbers of species. The arithmetic of extinction does not apply only to species. Genera may disappear, followed by families.

Reason Two: The constant reminder of taxonomy reinforces the message that all living things, though they occupy specific compartments, are related through having descended from common ancestors. When successful in its classifications, taxonomy shows us the relatedness of living things. To know that crows and blue jays both belong to the family Corvidae brings a sudden smile of recognition. The cry of the blue jay is simply a high-frequency "caw." Both birds are excellent mimics, are successful scavengers, have thick, strong bills, and so on.

Reason Three: We must give science its due as the court of last resort in deciding what is happening to our natural environments and why. Taxonomy is the only key to knowing what species are actually in the process of disappearing. Vacuous estimates of worldwide extinction, based on island biogeography theory and extrapolated to the world at large, are a poor substitute for precise knowledge.

In this book, scientific names for species are italicized. Normally, they will consist of two parts, genus and species. Sometimes only the genus will be named, the species being either irrelevant or unknown. I have cast a few single names (genus or species) in less formal style by making them into characters, such as Lotor the raccoon and Cymbella the diatom. Here they are not italicized. In field guides and other biological publications, the common names of species are capitalized. Here, for the sake of typographic uniformity, they are not.

I have taken also a few liberties with present knowledge of natural places and their inhabitants. Though they are for the most part realistic, the portraits to come are occasionally speculative. The boldest assumptions are explained in notes at the back of the book, under the chapter titles.

Virtually everyone agrees that natural environments are disappearing at an alarming pace. The world's great tropical forests are being cut for agricultural land or for timber. Ninety percent of the habitable portions of eastern North America has become a permanent clearcut. Some large, fairly pristine blocks of forest remain, mostly in hilly or mountainous country. But the rest is much like Hungry Hollow: isolated patches of forest, field, and wetland, slowly and steadily bleeding species. The evils of pollution are well documented and undeniable, but the loss of habitat remains the main environmental threat at the millennium. In North America it proceeds forest by forest, pond by pond, with little or no notice from the media. As we nickel-and-dime the world to death, Hungry Hollow hangs in the balance.

Acknowledgments

Although I have published research in mathematical biology, although I am an active naturalist, avocational biologist, government ecological committee member, grower of native trees, amateur microscopist, and biological research team member, and although I have been taking courses in biology, I am not, alas, a qualified biologist. Nor am I a chemist or a physicist, though I have often wished I were.

Even if I had been any or all of these things, I would still have asked scientific friends and colleagues to go over these chapters for errors and omissions. I thank them sincerely and readily acknowledge that any faults that remain are my own. For reviewing the manuscript, I am especially grateful to the following people: Dr. Graham Cairns-Smith, Chemistry, *The University of Glasgow;* Dr. Stan Caveney, Zoology, *The University of Western Ontario;* Dr. William G. Hopkins, Plant Sciences, *The University of Western Ontario;* Dr. Bryce Kendrick, Biology, *The University of Victoria;* Dr. Bill Taylor, Biology, *The University of Waterloo.*

I acknowledge with gratitude the assistance of David Maracle of the University of Western Ontario's Native Languages Center.

I must also thank Anita Caveney for a careful reading, as well as William Frucht, Steve Pisano, Theresa Shields, and Connie Day for skilled editing assistance. The artists, Christie Lyons and Roman Szolkowski, have become partners in the vision of Hungry Hollow and, finally, I thank Jerry Lyons (no relation), publisher of Copernicus, and my agent, Linda McKnight, who share our vision.

A.K. Dewdney

Hungry
Hollow

Procyon lotor

I t is February, during a thaw. The old raccoon Lotor leaves his hole in the hackberry tree, clambers out a bare branch, and gazes at the fading western sky. Behind him, the starry wheel of night turns imperceptibly, renewing itself. Procyon, coldly brilliant, has just risen in the east behind him. The name means "before the dog." Soon Canis Major will come sniffing over the horizon. It will leap in slow motion through the hours of arc in endless pursuit of Procyon.

Lotor's fate is recorded in the horoscope of this night. The ancient chase of Canis after Procyon has been enacted many times in Lotor's life. If there are still raccoons, it is because Lotor's family, the procyonids, had a head start. His family came "before the dog."

The procyonids, represented today by the raccoon (*Procyon lotor*), ringtail, coati mundi, kinkajou, cacomistle, and red panda, diverged from the carnivore line millions of years before the canids. The wild dogs grew fast and sleek, hunting in

"Lotor's fate is recorded in the horoscope of this night."

packs, while raccoons remained slow and thickly furred, preferring to hunt alone. The reason is simple: Dogs were committed to eating meat; raccoons were not. Pack life made it possible to hunt the larger game that dogs needed. But solitary life was sufficient—even necessary—for securing the raccoon's widely varied diet of smaller items.

If the ancestral carnivore was not omnivorous, the procyonids stumbled onto the art themselves, eating not just flesh but also fruits and nuts in season. That is how Lotor survives today. His smorgasbord is immense but sparsely laid out: a grub here, a berry there, and over by the creek a frog. Lotor is hunter, gatherer, and scavenger, a true generalist when it comes to food.

His main duty is to himself, to survive. But he has another duty, a purely symbolic office. He is *spiritus locarum* for Hungry Hollow. He is main character, sometime guide, and witness to everything.

Lotor's badge of office is the antique mask he wears. Did the most ancient carnivore wear such a mask? Lotor and the red panda have one. So do the giant panda of the bear family and that antique canid, the raccoon dog of Central Asia. And so does the African palm civet. The oldest mammalian eyes stare out from these masks. Lotor's are shiny, black, and hard-looking. The procyonids have seen a lot.

Humans assume that the mask marks a thief. Without doubt, Lotor and his kind are outlaws. They ignore our legal system, filching food at every opportunity. But from the procyonid point of view, there is no moral distinction between an open kitchen window and a stream bed free of ice. Food is where you find it.

Procyon's real crime is committed only rarely. To make off with a human heart propels him to the pinnacle of environmental mastery. He gives up petty theft altogether, for the human feeds him on demand. Yet in the midst of his procyonid paradise, the raccoon does not become subservient like the dog. Nor does he pretend to be independent like his more distant carnivore cousin, the house cat. Raccoons have never been standard pets.

A mere million years ago, some wild dogs grew too fond of the new campfires and of the hominids that sat around them. They became subservient, watching over the camp in exchange for scraps. Eventually the humans bred them into the mutant forms that whine and bark in the night of the encroaching suburb. "That's what happens," says Lotor in his spiritual capacity, "when you lose your independence." For fear of the dogs, Lotor has yet to enter the suburb.

—— ——

It is a warm July evening. The setting sun burnishes a hackberry near the edge of the forest of Hungry Hollow. Far up the trunk, inside a hole, two eyes glisten behind their mask. The old raccoon yawns, airing his sharp white teeth. He pokes his head slowly out of the hole and peers through the screen of trees. He can see the floodplain meadow, Hungry Creek, and the cliff above it. He will dine on crayfish and snails in the coming darkness—if the dogs don't come.

Lotor knows where they live. Last spring, he waded Hungry Creek, climbed the steep bank, and then crept cautiously through the strip of remnant forest at the top until he came to a clearing where he could study the lights, the sounds, and (best of all) the odors. He saw the back of a house and an opening where a head passed back and forth. A delicious smell came from the house and his stomach rumbled. Scents drew him forward, but noises warned him back. The sounds of screen doors slamming and children screeching spoke of a certain violence, although he did not know what made them. But he knew the baying and barking. Whatever else it might be, the suburb was a place where dogs gather.

The little brown bat squeaks as it wheels past Lotor's home. By the light of echoes, its ears see the mosquitoes that swarm around Lotor's hole, and the bat picks them off, one by one. The song sparrow sings once more to the setting sun, then settles into its nest among the hawthorns and crab apples by the creek. A sawyer beetle grinds its last sawdust of the day with a faint ticking noise in a dead limb above Lotor. And throughout the meadow below, the field crickets intensify their chorus.

It is time. Lotor climbs out of the hole to one side, reaching for the ridges of bark, swinging his tough old body onto the trunk.

He used to go down headfirst, but these days he prefers the safer backward descent. Unhurried, he feels his way down the trunk, the ridges as familiar as stair treads. When he looks over his shoulder, the meadow comes up to him by stages. On the way, Lotor pauses occasionally to watch, listen, and sniff.

The evening air is a potpourri of scents from near and far. Blue-eyed grass, red clover, common milkweed, tall meadow rue, dame's rocket, wild geranium, and a hundred other species pour fragrances strong and faint into the air. Each flower, leaf, and gland lays faint traces in the twilight. The green factories of photosynthesis have closed down, and every plant respires, living on the starches accumulated in the light of day.

Earth, leaves, and decay blend into a musky bouquet from the base of the tree. And Hungry Creek across the meadow is awash with the smells of its own digestion. What is there to eat? Lotor moves slowly toward the creek, pausing here and there to sniff. He makes a slight whiffing noise as he goes.

Grubs can be scented in the soil and in dead logs. Holes in the ground, large and small, need checking. The smell of ants is acrid and distinctive. The earthworm's burrow has a pleasant, meaty smell, and Lotor bounds forward, front legs moving together like crutches. He feels with sensitive paws. They touch a worm and close quickly before it can withdraw. Lotor brings it wriggling to his mouth and downs it like a large noodle of spaghetti. It is delicious! Further on, he comes to the burrow of a garter snake. The hole is rank with dead skin. Close to the creek now, he passes a groundhog's hole by the roots of a sycamore tree. The burrow has a sweaty, not unpleasant smell.

He arrives at the chuckling of Hungry Creek, downstream from the bend. Here is a wild plum tree, its fruit already beginning to fall. Lotor pauses, snuffling through the reed canary grass and spotted Joe-Pye weed for the soft green treasures. Finding one, he carries it off in his mouth. He looks guilty but does not feel at fault as he glances warily upstream and down. Then he dips the fruit in the shallow, barely flowing water, washing it in two paws, small black hands, until it is thoroughly wet. His name, Lotor, means "washer." He hunches forward to seize the fruit in the side of his mouth, sharp carnassial teeth slicing flesh from the stone. He swallows the sweet mass and then goes back and eats another plum. And then another. He has barely started.

Lotor works his way slowly upstream, reviewing each landmark in his mind. Everything must look and smell and sound the same, or he will not proceed. Here is the stone bed, an abandoned watercourse where orchard grass grows in clumps. Halfway across, Lotor's nose picks up something new. This is special. But now it is gone. Lotor scuttles to the water's edge, rears up like a miniature bear, and smells again. It is coming from across the creek. Without hesitation, he wades noisily into the water and crosses, half swimming, to the other side. Here is a path, and here also is a piece of plastic and the source of that intriguing smell.

He picks up the piece of sandwich and carries it to the water, where he eats some of it before the ablution, and then some more. Lotor has never before tasted peanut butter and jam. He can smell nothing but peanut butter and jam. Hungrily he explores the entire area, from the water's edge to the path and beyond, where the bank grows steep and trees can barely find a roothold. He licks the plastic and then takes it to the creek to wash it.

The sun has slipped below the steep cliffs above Lotor, and a crescent moon is rising behind him. In the rapidly gathering darkness, a nighthawk screebs above and the great horned owl asks, "Who-whooo?" Lotor's blood quickens. Something about wading in the shallows makes him forget the sandwich. He is the hunter and his ancient blood is astir.

He trundles quickly upstream along the shore until he comes to the shallows at the bend of Hungry Creek. Opposite is the steep cliff of shale and limestone where the current has cut a deep channel. In the moonshade of black willows, Lotor plies his trade on the muddy bottom. He feels here and there, supporting himself first on one paw, then on the other. He works his way forward, following the shore with a slow groping until something subtle moves beside his paw. His arm darts and his eyes look abstractly up at the stars. He has a crayfish. It vainly arches to catch this grip in a grip of its own. Lotor does not wash the crayfish. He carries it to shore, pins the claws against a rock, and eats the tail first, smacking his jaws. The juices of the crayfish blend with the creek as they drain down his throat. Lotor is happy.

Something descends the bank across the stream. It rolls and tumbles with a wet noise, then plops into the creek. Lotor pauses only a moment before deciding it is not an animal—at least not one for him to worry about.

A fossil has become dislodged by the water that seeps from a limestone ledge halfway up the cliff. The mineralized body of a long-extinct line rests on the bottom of Hungry Creek. A brook stickleback noses it briefly, then slips quietly away. It is the trilobite called *Phacops rana*. If it were still alive, Lotor would find it every bit as tasty as a crayfish. Behind Lotor, deep in the forest, a screech owl calls. The ghastly descending whinny hints at the ghosts of long-departed creatures and the tragic passing of an ancient environment.

The Tippecanoe Sea

O n the last day of its life some 390 million years ago, the trilobite called Rana scuttled across the clay and silt bottom of a shallow tropical sea, the Tippecanoe. There was nothing to distinguish Rana's locale from the sea bottom a hundred or even a thousand meters away. But it may be called Hungry Hollow because one day it would be.

It was late in the Devonian period, and North America was barely recognizable. The Tippecanoe Sea filled its interior, reducing the continental map to an enormous horseshoe open to a worldwide ocean in the south. A broad upland formed the western arm of the horseshoe, and a narrow mountainous strip nearly the width of New Jersey formed the eastern arm. The continent also tilted about 30 degrees, its northeastern margin jammed tight against a proto-Europe. The equator crossed the Tippecanoe Sea not far from Hungry Hollow. The climate was tropical.

The slow but immense force of this collision had created the chain of mountains called the Acadian. Himalayan in grandeur, this snow-capped range dominated the east coast of North America. It is still with us, eroded to a chain of humps in eastern Appalachia.

To the west of the Acadian mountains, the shoreline of the vast Tippecanoe Sea ran down the west side of what is now Quebec, New York, New Jersey, and the Carolinas. The sea stretched across two-thirds of the continent. Rivers great and small drained the eastern mountains, coursing through fields and low forests of alien-looking plants in the Acadian lands. Here were lowly mosses and liverworts, ferns, and clubmosses growing in the shade of tree-like seedferns and cycads. There were spiders, dragonflies, cockroaches, mantids, and other arthropods. There were amphibious animals, some frog-like, some lizard-like, that wriggled through the shallows or crawled on land.

The rivers fed sediments—clay and silt—into the sea. One of the rivers discharged a vast, nearly invisible plume of sediment that slowly settled out of the water as it drifted farther and farther from shore. Currents carried the plume a hundred kilometers to the west, where it passed over Hungry Hollow. A continuing rain of particles too small to see slowly built the soft bed on which the trilobite scuttled. Rana gazed at a varied scene lit by a fierce sun that ten meters of water overhead could hardly dim.

A hundred years earlier, there had been a large, flat, shelving reef here, but it now lay buried beneath two meters of sediment from the distant river. The reef building had ended for now, but the shelf would one day re-emerge as the lower ledge in the cliff above Hungry Creek.

On the last day of its life, the trilobite did not die in the jaws of a predator. Nor did it gently give up the ghost after a long and fruitful life of scavenging. That morning there had been an earthquake, another grinding tremor from the ongoing collision with Europe. The bottom had rocked momentarily, and the wave of compression that raced through the sea floor kicked up a small cloud of clay that gradually settled again. Rana still had some time.

Its eyes embodied a Greek and Latin prescription for survival. Rana lived by a multitude of extraordinary lenses (*phaco*), hexagonally defining two windshield eyes (*ops*). These bulging digital scanners resembled the eyes of a frog (*rana*), their overall shape implying the same burial tactic: Hide the body in one medium right up to eyes that would be mere bumps in the other medium.

With such eyes Rana peered into the sunlit waters, playing the old game of predator and prey. It was not buried at the moment but moved furtively in short dashes. Underfoot, the soft bottom yielded easily to the 30 jointed limbs that fanned centipede-like from beneath the chitin/protein armor. Under Rana's head, two large paddle-like appendages waved ceaselessly, fanning bottom debris and tiny swimmers into a mouth that was hardly more than a pore. It fed much as Lotor's crayfish would feed 390 million years later.

As the trilobite waved these coxae, wafting food to its mouth, its frog-like compound eyes kept constant watch on the ever-shifting scenery. All around it, horn corals sprouted, solidly sunk in clay, dense colonies of transparent polyps crowding their tops. Great bryozoan fans barely waved in the gentle, benthic currents. Feathery crinoids, sea-lilies with ropy arms, kept the slow rhythm.

The trilobite scuttled over a brachiopod, a lampshell that pulsed open and shut, feeding on what the currents might bring. Rana stumbled (though with only a few of its jointed limbs) on a small snail crawling over the brachiopod's shell.

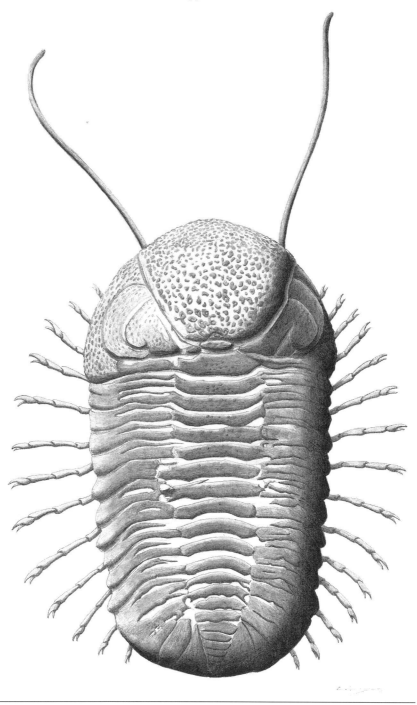

"With such eyes Rana peered into the sunlit waters."

The snail grazed on the field of green algae coating its fellow mollusc. When Rana had cleared the lampshell, the gentle current brought a dead flatworm slowly into the sensorium. Dinner gongs rang in the trilobite's nervous system as eyes signaled the prize and sensitive antennae touched it. Rana grasped the flatworm to its mouth, fanning the tasty, decayed particles of flesh toward the pore, sucking them in. Blit, blit.

This was the ancient life of Hungry Hollow, from crinoids and trilobites down to the tiniest protozoa, algae, and bacteria. The big forms of that time are no longer with us, but the small ones are. They had already been around for more than a billion years with very little change. It was a typical living community—some large creatures, some small, some common, some rare, some already ancient, some recently evolved.

Rana's eyes detected a larger form moving behind the bryozoan fans, suspended among them. Its armored head probed the muddy bottom, and a dozen shapes, small and strange, fled with jerking and darting motions. The placoderm fish, plates flashing in the shallow sun, was hunting for trilobites. Rana fanned its 30 limbs unobtrusively and gradually dug into the soft bottom, sinking out of sight until only the frog-like profile remained. To see and not be seen.

The placoderm, finding no trilobites around the base of the bryozoans, nosed its way toward Rana. The trilobite dug deeper, settling until even the eyes were covered by a fine sediment. They had just seen their last. Before the placoderm could even guess what lay beneath the two bumps in the bottom, all was suddenly confusion and darkness. A tsunami struck.

Shock waves of the morning's earthquake had also passed under much deeper water in the Tippecanoe Sea. The bottom had heaved up and down by a meter or so, moving untold tonnes of water above it and creating a large but subtle wave, barely noticeable at the surface. It was at first a gentle swell hundreds of meters in length. But as it moved toward the eastern shore of the Tippecanoe Sea, it fell subject to an age-old law: When the depth of water beneath a wave decreases to half the length of the wave, the energy within can no longer be contained in a gentle, sinusoidal form. It will begin to break and rear up from the sea, a thing alive.

A huge roller formed as the tsunami raced toward the Acadian shore. At Hungry Hollow, the angry brown behemoth with a crest ten meters in the air devoured everything in its path. A huge sucking plucked the bottom like a rug, unraveling it into a cloud of mud, clay, brachiopods, placoderms, and trilobites, all tumbling, turning, and finally settling in a rain of debris after the tsunami had passed. Everything was buried except a few of the stronger armored fish that could wriggle free in time. Rana was not so lucky. Under nearly a meter of muck, it beat its legs in futility, then slowly lapsed into death as it ran out of oxygen.

The earthquake that had created the tsunami also made less radical but telling changes in the landscape of Acadia, lifting some areas up and lowering others. The course of the river changed; it now poured into a swamp twenty kilometers south of its previous mouth. The new plume missed the westward current. The faint but steady supply of clay and silt to Hungry Hollow ceased. Gradually, the reef-building algae returned to the area, no longer to be buried. Within a few decades, a new shelf had begun to form. It would ultimately equal the lower shelf in thickness, and the clay and silt sandwiched between would harden over the eons to shale. Geologists would call it the Hungry Hollow formation.

The story is still there in the rock above Hungry Creek. A fossil community of once-living organisms, some forms intact, others mere smudges, lie frozen at the far end of a bridge over time itself. By a supreme irony, the fossil Rana has landed upright at the bottom of Hungry Creek. Tomorrow its eyes will gaze once again into sunlit waters, staring blankly at forms more bizarre than itself.

Dianne sapiens

Above the cliffs and behind a strip of remnant forest, there is a community of sophisticated primates. They walk on their hind legs and use their front legs for grasping and carrying things. They are inquisitive like Lotor, but their judgment is less well developed. The primates have language and call themselves "human beings" or, sometimes, *Homo sapiens*. Being very good with their front paws, they have constructed large, box-like structures called "houses" in which to live.

It has been 390 million years since the Hungry Hollow formation was laid down. If those 390 million years were compressed into one, starting a year ago, then Lotor's kind would have been around for the last month, human beings for only a few days. It would be 15 minutes since the end of the last ice age and 7 minutes since *Homo sapiens* developed agriculture and began to record its own history. The Industrial Revolution would have begun a mere 15 seconds ago.

The abundance of this species has grown considerably, from a few million a thousand years ago to several billion now. The sapient primates live in every conceivable place. Everywhere on the planet, farms and cities have elbowed their way through natural environments until the original homelands of *Homo sapiens* have been all but erased.

One of the new primates calls herself "Dianne." She belongs to the colony and lives in a house that backs onto the ridge of trees above Hungry Creek. From her kitchen window she can look through a gap in the trees. She can see the meadow and a bit of the forest behind it.

The colony is called "Whispering Pines, a Concept Community." As far as Dianne can tell, the "concept" involves only the pines, there being nothing else to distinguish this colony from a thousand others. The pines have been planted sparsely throughout the small development, but now, for reasons that Dianne has considered looking into, they are all dying. A wildly curving road divides the concept community into a jigsaw puzzle. It twists by identical houses and lawns. It turns by identical pines, all identically turning brown.

Until recently, Dianne taught biology at the university. Since last January, when she learned of her impending termination, she has been looking for work, a job in which she could use her biological skills. Her specialty is microbiology, the study of the tiniest living organisms. On the first of July, keeping her emotions sternly in check, she brought home the microscope from her student days. She also brought home glassware and a few chemicals. She set up a small laboratory in her house.

Since then she has been doing water analysis for the county and for whatever other clients she can find. She wants desperately to keep her house, symbol of the independence she has worked for. She does not yet know it, but she cannot leave Hungry Hollow. She is the *Homo sapiens* of the place.

She has lived beside Hungry Hollow for five years, and almost every day, in warm weather or cold, she has walked along the creek, across the meadow, or through the woods. Until recently, she used the Hollow as a refuge, a natural backdrop for thoughts academic or personal. Every now and then she would stumble upon a plant with white berries or a mushroom with a red cap, not knowing what they were. On several occasions she stood on the bank of Hungry Creek, staring at the olive-colored bottom and pondering the gulf between her research and the "wild material" abounding in the creek. There was a biofilm there, she knew, churning in its billions. Few microbiologists had ever studied such environments closely.

These days she has more time. Her teaching has come to an end. Without access to an electron microscope and other high-tech equipment, her research languishes. Although she does not realize it yet, her life has been opening up, and when she walks the Hollow, it is like a new world. She pays attention. What bird is that? She buys a bird book. She would know the red mushroom and the plant with the white berries, too.

As she disengages her self-image from the world of academe, she reviews the ideas she once accepted without a thought and finds them wanting. Why do biologists insist on calling natural environments "systems"? As if they were some kind of machinery in which cogs meshed and valves opened and closed at regular intervals. Why do they refer to the plants and animals that live in the same place as a "community"? As if each knew the others and all took their places in the food web, cogs in the ecosystem. Other standard ideas look increasingly like fantasies to Dianne: competitive exclusion, plant succession, ecological niche, even biodiversity itself. She has been mildly shocked to discover that there is no general agreement on what the word *biodiversity* means, let alone on how to measure it.

Dianne is fearful at such thoughts, like a nun who questions dogma. She sniffs at the ceiling, horrified at how angry she has become. With what will she replace her old world?

It has grown late. She goes to the kitchen, where in spite of her better judgment, she pours a cup of coffee. She looks out into the blackness beyond her window and sips the bitter liquid. On a whim, she puts a few cookies in the pocket of her sweater and steps out into the warm summer night.

There is a path down the ridge to the east side of Hungry Creek. The moon paints the way in patches of ghostly light. She has become nocturnal, part of the night. Her thoughts are no longer rational but primordial. In spite of an inborn caution, she walks easily and without fear, feeling her way with new senses.

She sits on a smooth granite boulder beside the creek and listens to the water. She is perfectly silent. Slowly the place assembles itself from the shadows. It is not an ecosystem, she decides, but simply a natural place. It has a meaning and the meaning is part of her. The night, the mystery, the beauty—all belong to her. Natural places belong only to those who visit them. She forgets to eat the snack in her pocket.

There is a stirring by the shallows across the creek. Something moves in the shadow of a young willow, and Dianne catches her breath, all ears and eyes like an animal, *Homo sapiens* of old. There is no mistaking the form that emerges into the moonlight. It is a raccoon. It seems unaware of her as it paces the shore. It reads the bottom for hidden dangers such as the very large snapping turtle, *Chelydra serpentina*. The raccoon stops frequently, raising its sensitive nose in the air, sniffing at something elusive but interesting. When it finally enters the water, she watches it feel with its paws, sweeping, then placing them firmly to trap what might be trapped.

What is the raccoon hunting? There may be frogs there, or crayfish, she thinks, but she never finds out. The raccoon spends a timeless five minutes searching the bottom before it regains the shore to scratch at a flea on its shoulder. "You are a big old boy, aren't you?" says Dianne under her breath. She doesn't know whether it is old or a boy, but she is right on both counts. Is it the moonlight?

She feels sorry for the raccoon and instantly remembers the cookies in her pocket. How dare she? One should never feed wild animals. But the night, the Hollow, and the raccoon all belong to her and she to them. She will follow her instincts. Silently, she withdraws a cookie from her pocket, takes careful aim, and throws it toward the animal. It lands beside him and he is gone instantly into the bushes. Dianne feels disappointed, then foolish. That's what you get for for disobeying the rules.

The spell is broken. She finds herself sitting on a rock by a nondescript stream in the middle of nowhere at night. She is about to get up and leave when she hears a nervous, chuffing noise. The raccoon re-emerges from the shadows, drawn by the powerful scent of the cookie. He scans the far bank but fails to see Dianne and makes directly for the cookie, hungrily chewing half of it down. Dianne is enchanted all over again. As though to repay her with a minor miracle of his own, the raccoon picks up the cookie between his front paws, stands on his hind legs, and takes three ungainly steps into the water, where he falls forward, releasing his prize with a splash. He washes the cookie, catching the pieces with his paws, hand to mouth like a human. Dianne cannot help herself. She says "Hello" out loud. And in an instant, Lotor is gone.

As she climbs the ridge back to her house, she feels happy and whole. She does not realize it, but a new life has just begun. Lotor, it must be said, has no such feelings. He is terrified of the animal that throws food in the dark. It is hostile and, because it throws food away, obviously insane.

The Hackberry

The hackberry tree in which Lotor sleeps is immensely strong and heavy. In the large, its strength derives from the structure of annual cylinders, nested each within the next like Russian dolls. In the small, its strength depends on a polymer called lignin, the essence of wood. These strengths, structural and molecular, are common to all trees: root, trunk, and branch.

The hackberry tree is massive with the tons of carbon dioxide it has absorbed from the air. It has taken these molecules apart, rearranging the atoms of carbon and oxygen into organic molecules. All but a small fraction of its mass has literally materialized out of thin air.

The weakness of this hackberry can hardly be suspected at first glance. The sturdy-looking trunk and limbs conceal a building full of apartments and restaurants, home and host to thousands of tenants and visitors. Its chief resident is Lotor the raccoon. Altogether, these occupants will eventually bring the old tree down.

The hackberry began with a bird-borne seed that germinated one spring after lying under leaves over the winter. A year later it had grown to Lotor's height, consolidating its gains in lignin. In the following year the lignified part did not grow. Instead, the woody seedling wrapped itself in a new layer of soft tissue while the bud at its tip sprouted, extending by cell division and elongation to double the height. The sapling's new outer layer and shoot lignified in turn, freezing its form again. In its third year the young hackberry laid down another annual layer of fresh outer wood and, at its tip, grew another shoot. The woody parts did not lengthen, nor would they ever. Only at branch buds did the tree make new extensions each spring, and only under its bark did it grow. Outward.

Today the hackberry's first seedling still stands, locked in the heart of the tree at its very base. The second-year sapling surrounds it now, as it did then. The branches these saplings once made have disappeared, stripped by deer or damage, present only as abortive pegs protruding outward through one or two cylinders. If the hackberry were cut across, the cylinders would appear as annual rings. If the tree were cut lengthwise, the buried branches would show up as knots. The solid heartwood kept cylindrical records of all the hackberry's former selves, one for each of its 56 years.

The life of the hackberry dwells in the outer layer, just under the bark. A thin cylinder called cambium, no thicker than paper, generates all outward growth of the tree. Every year it makes the long, tough xylem cells of wood just inside itself, stretching outward to accommodate them. Every year it also makes the shorter cells of living phloem just outside itself. Older phloem, pushed ever outward, dies and turns into cork.

As the tree expands, the corky layer cracks into pieces that cling as bark, protecting the living tissues just inside.

In the hackberry building, the xylem bricks also serve as plumbing. The wood cells of the tree are tiny pillars of strength. But they have lost their end-caps and now join together, making long pipes that extend from root to branch. Five thousand thirsty leaves suck on these microscopic supply tubes, their combined demand drawing water up the wood. The leaves of the hackberry need this plumbing in order to run the astonishing machinery of photosynthesis that, from water, carbon dioxide, and sunlight, manufactures carbohydrate chains—food for the tree.

The sun's energy, now stored in the bonds of carbohydrates, enters the tree's tissues through a separate plumbing system. The phloem cells just under the bark use pipes of a different kind. Holes in their end-caps permit the carbohydrate solution to filter from cell to cell—a system that feeds branches, trunk, and roots as well. The 56 inner cylinders of wood bring water up, and the single outer cylinder of phloem brings food down. Between them, the nearly invisible sheet of cambium hides the secret of life and growth. In these things all trees are equal.

Even as the hackberry tree continues to grow, it has already started to die. It is under attack by its tenants and visitors. Insects, fungi, birds, and mammals contribute to the slow destruction. Lotor's den has grown so large that only 20 percent of the heartwood is left to conduct water upward during the warm weather.

Lotor's apartment began as a thick, lower branch on the hackberry. One evening twelve years ago, a windstorm proved too strong for the branch. Its cylinders of lignin bent and then shattered in the horizontal sweep of wind. The branch broke at the base, its splintered heart splaying out, unsightly, from the trunk. After the storm, breezes played through the hackberry leaves and left fungal spores, the omnipresent and invisible air force of decay.

Some of the spores came from wood-digesting fungi. These spores germinated among the fresh splinters at the branch's base. Tiny transparent tubes, the hyphae, insinuated themselves into the wood, secreting digestive enzymes as they grew. The magic liquids broke the lignin chains, making simpler, smaller molecules for the hyphae to absorb. Some insect larvae ate the enriched wood; others ate the hyphae.

Woodpeckers busied themselves around the damaged limb, prying away the splinters and chips of rotten wood with their powerful bills and exposing fresh surfaces. There was great slaughter among the grubs, not to mention the many adult insects that lived in burrows and crevices of the wood. But the ever-growing molds cared nothing about the slaughter, nor did the insects that survived.

The patch of rotten wood at the trunk's surface gradually deepened into a cavity. From there its first tenant, a saw-whet owl, kept nightly vigil over the meadow one summer, watching for small creepings in the goldenrod and Queen Anne's lace. Then came two mourning doves to make a precarious nest the following year. Squirrels raised a brood in the enlarged hole a year later. Then came the young Lotor, who was not above scrabbling at the soft walls of the hole with long, hard nails. At first there was barely room to curl up. Lotor used the apartment only for the summer, spending the following winter under a brush pile deep in the woods.

The hole that Lotor calls home smells dank and crawls with several dozen species of the small, chitin-armored beasts called arthropods, the joint-legged ones. They include isopods such as wood lice, chilopods such as centipedes, arachnids such as spiders, and insects, including Lotor's own lice and fleas. The insects are most numerous. There are beetles, ants, flies, wasps, earwigs, moths, and, just outside the hole, a cicada emerging from its formal nymph-wear after seventeen years underground. In most cases it is the larval generation that has the grand appetite, from the cicada grub that nibbles the hackberry's roots to the beetle larva that chews on the half-rotted wood lining Lotor's snug den. The tree has fought back, growing around the edge of the hole, furnishing it with a smooth collar of new wood. But it will never succeed in closing the old wound.

Other apartments have more specialized architectures, each made by just one or two species of tenants. Not surprisingly, they are all arthropods. The leaves of the hackberry are studded with green nipples, each one housing the minute green larva of a tiny plant louse with an enormous name, *Pachypsilla celtidismama.* The larva hatches in mid-leaf and then secretes a hormone that causes a new growth in the surrounding plant cells. A gall erupts from the surface of the leaf, a fleshy swelling with a small room in the center. It has no door or windows. The larva feeds on the walls. The hackberry nipple gall louse, as the name implies, cannot live on any other tree.

On every branch of the hackberry, a benign symbiosis has created hundreds of witches' brooms. A species of the gall mite *Eriophyes,* a tiny arachnid, lives in the branch buds. It carries with it a fungus, the powdery mildew known as *Sphaerotheca phytophila.* Spores of this fungus germinate near the buds. *Sphaerotheca,* in turn, produces a chemical that stimulates the plant to produce many short sprouts (the witches' brooms), instead of a single long twig. Within this cluster, *Eriophyes* finds its hunting multiplied a hundredfold. Like the nipple gall louse, it has committed itself to the hackberry.

The co-habitation, not to say marriage, of insect and fungus is common in trees. The female ambrosia beetle *Monarthrum* has made extensive apartments in a dead branch high in the hackberry. Into the heartwood she has excavated an elaborate tunnel. The job was made easy by the fungus *Ambrosiella,* a species of which the beetle brought with her on her first visit to the old limb. She carried its spores beneath her jaws. The spores germinated in the fresh wood, penetrated its tissues, and in time produced new hyphae that the beetle could eat. The beetle chewed new passages in the softer, decayed wood, tending the fungal walls and dining continually. She made special larval nursery rooms for her offspring, and when they hatched, they were free to enlarge their mother's building plan or to seek their fortune in a wider world.

Dianne has passed under the hackberry just once, but did not notice Lotor's apartment far overhead. Instead, she stumbled on a large piece of bark that had fallen from the dead limb. She was intrigued by the appearance of an extensive script engraved in the wood. The secret borings of the bark beetle, *Scolytus,* formed a pattern of galleries. The mother bark beetle made a winding main passage, and her numerous offspring made smaller passages sprouting from it. The pattern made Dianne think of an ancient, long-forgotten message. In her new life she is free to imagine what it might say.

The hackberry may be Lotor's home, but he leaves it for the evening shift. Other animals must remain there day and night. For them the hackberry is more than home. It is a world, and the world has five kingdoms.

In Kingdom Animalia, the animals, there are Lotor and his co-tenants. There are the birds that come and go in the branches. There is the nest of carpenter ants at the base of the tree. And behind the bark eel-like animals called nematodes, too small to see, twine themselves restlessly within damp pockets, eating still tinier lives.

The molds that will ultimately be most responsible for the old hackberry's downfall, from the beetle's ambrosia to a rubbery polypore that brackets the trunk, all belong to the Kingdom Fungi.

Growing in minute cracks of the bark are one-celled algae such as *Chlorella,* microscopic green spheres that form large colonies wherever there is a damp foothold bathed in sunlight. They make a greenish discoloration. Amoebae and other protozoa creep everywhere that water minutely clings, from the roots of the tree to its crown. *Chlorella,* amoebae, and the rest all belong to Kingdom Protista.

The tiniest independent life-forms of all, the bacteria, are ubiquitous. Take away the tree, bark, leaves, phloem, cambium, and wood (that is, remove everything *but* the bacteria), and a grey tree-ghost remains. The bacteria live on the complex chemical remains of practically every living thing in the hackberry, including the fungi. They belong to Kingdom Monera.

The remaining kingdom, Kingdom Plantae, is represented by the hackberry itself, not to mention some moss growing on the base of the trunk.

The hackberry is thus a microcosm of Hungry Hollow, itself a microcosm of Earth.

The Ant's Journey

Every year in Hungry Hollow, there is one magic day on which anything might happen. Today is the day. It begins with an amazing coincidence. The hackberry tree in which Lotor is fast asleep happens to match perfectly the Tree of Life, branch for branch and twig for twig (including the witches' brooms).

The Tree of Life for Hungry Hollow has roots in the past and leaves in the present. Its branches portray the relationships among living things. It is also a taxonomic tree, a classification of living forms into groups and subgroups. Humans have made it but have not made it up. The trunk is called Life, and from it extend five main branches, one for each kingdom, from bacteria to plants.

Monera   Protista   Animalia   Fungi   Plantae

So it is that the hackberry's trunk divides, within the space of a few meters, into five enormous kingdom-branches, some leaning outward, some straining upward. The five kingdoms branch

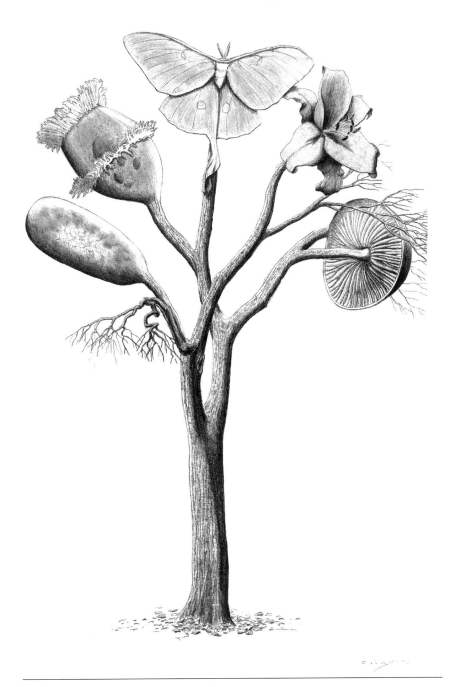

"The Tree of Life . . . has roots in the past and leaves in the present."

anew into phylum-branches, then into classes, orders, families, genera, and, at the end, into species-twigs, each bearing a leaf. On this magical day, the hackberry has one leaf for each species of living thing in Hungry Hollow.

The ants called *Camponotus pennsylvanicus,* otherwise known as carpenter ants, have made a nest in the base of the hackberry. A thousand sisters live and work there, subjects of a single enormous queen who lives in royal chambers at the center of the nest. Together, the carpenter ants have gnawed out galleries and brood chambers, depositing the sawdust in a heap outside the entrance beside one of the hackberry's root-buttresses. This morning, one of the sisters will climb the tree to its very top, a grueling journey, to search for aphids on the upper leaves. With luck she will find a herd and milk them for honeydew, lapping up as much of the precious fluid as her crop will hold. Then she will descend the tree, another long journey, down to her home where she will disgorge the honeydew to her sisters or, perhaps, to a developing grub.

The carpenter ant is one centimeter long, and the upper-most hackberry leaves are fifteen meters above her head. She will climb some 1500 times her own length for the honeydew, a feat that a human could match in scale only by climbing fifteen Washington Monuments, one atop another.

There is no denying the ant's beauty on this magical mid-July morning. The sun gleams on her polished black head, wide at the back and narrowing at the front to a pair of large mandibles, sawteeth meeting in a vertical grimace. Her eyes, which she keeps clean with her forelegs, also glisten, a polished ebony honeycomb of facets. Inside the head is a veritable brain that responds and remembers. Her plans may be little more than programs, but her eagerness seems unmistakable.

She ascends the rough bark, keeping to the valleys between the ridges. Hooks on her legs engage corky crevices and cracks in the hackberry bark. Leg muscles churn inside the rigid exoskeleton, and the carpenter ant nimbles upward at a dizzying pace, taking the Washington Monuments at one flight of stairs per second.

Where the trunk of Life divides, the carpenter ant chooses the great branch for Kingdom Animalia. Before long she arrives at a series of new branches that give off in various directions. These are the eleven phyla of animals, from the vertebrates (Chordata) such as Lotor to the rotifers (Rotifera), tiny monsters that swim in Hungry Creek. Their names, Latin and Greek, hint at their natures:

Acanthocephala (*spiny-headed* worms, parasitic in chordate guts)

Annelida (worms segmented in *rings,* living in soil or water)

Arthropoda (*joint-legged* animals, living everywhere)

Chordata (the vertebrates of Hungry Hollow, with dorsal *chords*)

Gastrotricha (tiny aquatic creepers, *bellies* covered with *hairs*)

Mollusca (*soft*-bodied animals, most having hard shells)

Nematoda (*thread*-like creatures, too small to see, everywhere)

Nematomorpha (visible *threadform* worms, parasitic on insects)

Platyhelminthes (*flatworms* on the bottom of Hungry Creek)

Rotifera (microscopic animals that swim with *rotors borne* in front)

Tardigrada (tiny, *slow-walking* animals that live in damp soils)

Every animal of Hungry Hollow belongs to one of these phyla, and every one of these phyla has at least one member in Hungry Hollow. If English were the language of taxonomy, we might know these phyla as Spinyheads, Ringers, Jointlegs, Chordies, Footbellies, Softies, Threads, Threadforms, Flatworms, Rotorholders, and Slowgoers.

The ant does not choose the branch Acanthocephala or the branch Annelida. Instead, she crawls quickly out the branch called Arthropoda, the joint-legged animals, from crayfish to insects to mites.

On the phylum-branch Arthropoda, the carpenter ant encounters a new set of eleven branches for the classes of arthropods in Hungry Hollow. All have rigid, articulated exoskeletons. The number of class-branches hints at the explosive evolutionary radiations enjoyed by the arthropods from their beginnings hundreds of millions of years ago:

Amphipoda (small, shrimp-like swimmers in the creek)

Arachnida (spiders that spin and mites that creep)

Branchiopoda (water-fleas that dart in the shallows)

Chilopoda (centipedes that scurry in the forest litter)

Collembola (springtails that shoot away on a moment's notice)

Copepoda (one-eyed cyclops that swim in Hungry Creek)

Decapoda (crayfish that feed in the shallows)

Diplopoda (millipedes that march on a thousand feet)

Insecta (insects that crawl, fly, swim, or burrow)

Isopoda (pillbugs that curl up when disturbed)

Ostracoda (seed shrimps that swim in pulses)

Every arthropod of Hungry Hollow belongs to one of these classes, and every class has at least one member living in Hungry Hollow.

The carpenter ant passes by the first eight branches and crawls nimbly up the Insecta limb, which is heavy with species. Viewed from the ground, this branch seems to continue the main trunk as it soars almost vertically upward. The exponential explosion continues as the ant encounters a new series of no fewer than fourteen branches for the orders of insects that live in Hungry Hollow.

There is the Loyal Order of Siphonaptera, fleas that ride on every bird and mammal of the place. There is also the Royal Order of Hymenoptera, which includes the social insects such as bees, wasps, and ants, with their queens and courts. *Camponotus* herself belongs to this order-branch, but she does not follow it this morning. Instead she picks the branch called Homoptera, the order that includes cicadas and leafhoppers. What is she up to?

The carpenter ant crawls along the Homopteran branch and now begins to tire as she nears the end of her sugar reserves. She arrives at her next set of choices, the eleven families of homopteran insects that live in Hungry Hollow. These include the cicadas (Cicadidae), the leafhoppers (Cicadellidae), the spittlebugs (Cercopidae), and eight other families, including the Aphidae, cattle of the insect world.

Aphids spend most of their adult lives on a single plant, bush, or tree. On a stem or leaf of choice they adopt a fixed position, insert a proboscis (tube) through the cuticle, and suck juices from the phloem cells of veins. Some aphids, such as

Aphis pomi, secrete wax from two pipes, or cornicles, on their backs. They also secrete the liquid called honeydew from their anuses. Like all insects, they need protein, and plant juices have so little that aphids must drink a great deal, excreting the excess sugar. The carpenter ant has no qualms, as she rushes along the tiny *Aphis* branchlet to the *pomi* twig, about milking her cows from the rear. We might understand if we knew how good honeydew is.

On names

The use of classical Latin and Greek to name species puzzles many members of the species *Homo sapiens.* A subspecies known as *H. sapiens biologist* has made up the names for good reasons. Not only do they indicate "specific" qualities, such as being blue (*coeruleus*), being wise (*sapiens*), or having four tails (*quadricauda*), but they are also more easily accepted by other members of the species *Homo sapiens* (wise). We are wise enough to recognize that the extraordinary privilege of providing fixed names for living things belongs to no living language. At the same time, each name, from kingdom to species, has a detailed definition that biologists (or general readers) may consult in cases of confusion. Some species look awfully similar to each other. Who will distinguish a red mulberry (*Morus rubra*) from a white mulberry (*Morus alba*)?

Those without even a smattering of Latin or Greek might still recognize names that reflect the Latin heritage of English. The bluebottle fly, *Calliphora vomitoria,* throws up on its food. To be more exact, it pumps from its stomach digestive juices powerful enough to dissolve meat into a thick broth that can be lapped up. The grizzly bear, which has never visited Hungry Hollow, is named *Ursus arctos horribilis* for obvious reasons.

By the same magic that makes this day special, the taxo-nomic leaf for *Aphis pomi* just happens to have several of these very aphids on it. They feed contentedly on the juices that course through the veins of the leaves. The carpenter ant, who dwarfs her translucent green cattle, milks them patiently. The feelings of the aphids in this matter are unrecorded, but they cooperate. For one thing, with a large and powerful farmer like *Camponotus* around, they are protected from mur-derous ladybugs and stinging aphid wasps.

What qualifies an organism for membership in Hungry Hollow's Tree of Life? What about the groundhog who lives just south of the forest on the edge of a corn field? Or the evening grosbeak that spent an afternoon hunting for a good nesting site before moving on? Like an exclusive club, mem-bership in the Tree of Life must be carefully defined. Bound-aries in space and time are called for.

For the boundary in space, a detailed map of Hungry Hol-low will do. An imaginary line passes north between the sub-urb and the forest on the bluff. It meets the highway and then turns abruptly west, making the shoulder of the road the northern border. It turns south at the county road, enclosing the upland forest on the west side of the valley. When the line meets the farmer's field a half-kilometer south of the highway, it makes a 90-degree turn and follows the barbed-wire fence that protects crops of corn and soya-beans. And so over Hun-gry Creek and back to the beginning.

Living things inhabit not a two-dimensional map but a three-dimensional space. The boundary must be extended into the third dimension, up and down. There is air above Hungry Hollow, and there is soil below. Let the boundary of Hungry Hollow extend down ten meters into the damp, dark soil and up a hundred meters into the air. The invisible lines have suddenly expanded into invisible walls that encounter, as floor and ceiling, horizontal planes that neatly enclose the living space of Hungry Hollow in a very large box. Humans, even biologists, love boxes.

As for the boundary in time, do we allow membership to the woolly mammoths that once roamed the Hollow when it was covered by black spruce? Arbitrary but fixed, the time of Hungry Hollow begins a year ago and ends today. Any living thing that has visited the great invisible box within the last year, no matter how briefly, will have its leaf on the Tree of Life, the leaf of its species.

Last spring when the hawthorn and swamp rose bloomed along the creek, some warblers visited Hungry Hollow on their way north. The Tree of Life therefore has leaves for the black-sided blue warbler, the Cape May warbler, and the yellow warbler, among others. Almost every day of the summer, a turkey vulture violates Hungry Hollow airspace, looking for something dead or dying among the meadow plants or along the banks of the creek. Its average altitude of 75 meters (not to mention its occasional landings) qualifies the turkey vulture for membership. Nor do other residents spend all year in the Hollow. The great egret winters in Florida, and the killdeer escapes the cold in Louisiana. The black-capped chickadee, by contrast, flies to northern Ontario for the summer.

However you define it, the Tree of Life has been steadily shrinking for the last two hundred years. Like the Hackberry, it has some dead limbs. Soon it will have more.

The carpenter ant's crop is full. She breaks off her milking and descends the branches in reverse order, from *pomi* to *Aphis* to Aphidae and on down.

pomi

Aphis

Aphidae

Homoptera

Insecta

Arthropoda

Then she pauses. There is a commotion up on the branch Chordata. The sound of birds squabbling in the sub-branch Aves makes vibrations that the ant picks up through her feet. None of her business. She hurries on to Kingdom Animalia and down the venerable trunk of life itself to her waiting sisters.

Animalia

LIFE

Congress of Birds

The hackberry leaves shone golden green in the early morning light, but the magical glow of the Tree of Life had already begun to fade. The birds balanced uncomfortably, one for each species, on their conceptual twigs of the great branch Aves. They were holding their annual conference on the state of the environment.

At first, no one could hear anything. A cacophony of voices from every flight of life delayed the congress: the buzz of the black-capped chickadee, the caw of the crow, the chirp of the house sparrow, the coo of the mourning dove, the hoot of the great horned owl, the "killdeer" of the killdeer, the pic of the downy woodpecker, the rattle of the belted kingfisher, the scree of the red-tailed hawk, the screeb of the nighthawk, the squawk of the great blue heron, the trill of the Carolina wren, the wall of the horned grebe. All these and more perched on the Tree of Life, a duly elected representative for each species of bird.

The great horned owl, by its imposing presence and persistent hooting, finally called the meeting to order.

"I am a creature of contraries. You vacation in the south and I in the north. You are small, I am large. On the ground my walk is ungainly, in the air I am swift and silent. During the day I sleep, at night I watch and hunt. In the West I am thought wise, in the East I am the symbol of stupidity. I am the embracer of opposites and a reconciler of views. We are assembled to consider whether our world is under threat and, if so, to consider what measures, if any, we may take. Our proceedings will begin with the customary minute of silence for the passenger pigeons that once flocked here in their hundreds."

After a minute of silence, which many birds spent searching their feathers for lice, the great horned owl looked around the assembly, silently counting. "It seems there are 93 present. At our last meeting there were 97 of us and, at the meeting before that, 98. We are declining in numbers. Assuredly we are declining. Let this congress take the form of a trial. Because humans are most often blamed for our troubles, let someone be humanity's advocate and let others bring charges. Who will be humanity's advocate?"

The house sparrow spoke up in the cultured tones that its species had not quite lost since emigrating from England. "I am the house sparrow of the family Passeridae. Humans are wonderful creatures. Without them we wouldn't survive. In the winter, humans in the suburb put up feeders for all who eat seeds or suet. I stand ready to defend my benefactors."

The great horned owl stared unblinkingly around the assembly with his huge eyes. "Then who will bring the charges? I see an upraised wing. Speak according to procedure."

"... *the great horned owl looked around the assembly, silently counting.*"

"I am the American crow of the family Corvidae. I will bring no charges, nor will my cousin, the blue jay. Our family supports humanity against any and all charges of mischief or criminal neglect. We crows are fed by the tires of their cars, which flatten many an arrogant mammal into a tasty buffet."

"I see another wing. Will you bring charges?"

"I am the herring gull of the family Laridae. Where would we be without humans? Garbage dumps are the best thing since the glaciers melted. We too support the human party."

"Who, who, who then brings the charges?"

After a long silence, the red-tailed hawk spoke up. "I am the red-tailed hawk of the family Accip . . . Accipi . . . Accipitridae. I sp-speak therefore for the f-falcons and hawks and eagles, all who are obliged to p-pursue their food. Our problems b-b-began when the farmers hereabouts started spraying their crops with noxious chemicals. We . . ."

"Murderer! You ate my babies!" It was the killdeer who spoke.

"You are out of order," said the owl. "You may try separate charges against the red-tailed hawk at the end of this session. Although I see nothing particularly wrong with the predatory way of life, I am prepared to admit that some of you may have different feelings on the subject. The issue before us is humanity, not predation. Please continue."

"We b-began to lose our eggs. Their shells grew thin, and when we sat upon them to warm young lives, their shells c-c-cracked and the yolk spilled into our nests. Our numbers began to d-dwindle until the farmers stopped using the worst of the chemicals, the one they call D-D-DDT."

"You see, they are not so bad," began the house sparrow. The owl ruled her out of order.

"We st-still do not live as we used to," continued the red-tail. Now we have heavy metals, P-PCBs, and a host of other p-p-poisons that I, since my own nervous system was contaminated, cannot even begin to p-pronounce. We eat things that eat these p-poisons. Worse yet, we eat things that eat things that eat these p-poisons. For example, I recently ate a s-song sparrow that had just eaten a lady b-bug that had just eaten an aphid that had been s-sucking juices from s-soybeans sp-sprayed with herbicide. With each step, the t-toxins c-concentrate. We do not live long. Our livers are c-c-contaminated and our reprieve from DDT hardly m-matters any m-more. I hereby charge humanity with g-genocide."

"The charge is noted," said the owl. Is there anyone else?"

"I am the belted kingfisher of the family Alcedinidae. I also charge humanity with genocide. In evidence I cite the pollution in Hungry Creek. The number of fish has been declining for years. Something is poisoning them. Or perhaps something is poisoning the things they eat, things too small for me to see. Hasn't everyone noticed the smell?"

"It is excrement," screamed the blue jay, forgetting his alliance.

"It is gasoline," wailed the grebe.

"It is death," squawked the turkey vulture.

"Order. Order! There is time for one more speaker. Then we will vote."

It was a woodpecker who spoke. "I am the downy woodpecker of the family Picidae. Once there were eight downys living in Hungry Hollow forest and in the woods above the bluff. But now there are just two, my mate and I. When humans made the suburb, they cut all the trees above the cliff, leaving just a strip for aesthetic purposes. We cannot live on aesthetics. We need grubs and plenty of them."

This stirring little speech was met with loud applause and a variety of chirps and squeals from the flicker, the yellow-bellied sapsucker, and other members of the Picidae family, not to mention insectivores such as the nighthawk, the Eastern kingbird, the meadow lark, the tree swallow, the nuthatch, the veery, the wood thrush and even the sparrows.

"What is your charge, then?" demanded the great horned owl.

The downy woodpecker thought carefully a moment. It was not just the trees; it was everything. "It hardly matters, since I must soon leave Hungry Hollow. But let the charges be genocide and the destruction of natural property."

A quiet voice interrupted—a sad, low voice. "I am the mourning dove of the family Columbidae. I will tell you something of these humans. Yes, I visit the feeder, but I also watch. When the others fly away, I remain to look in the windows of their box-homes to see what sort of creatures they are.

"They do not work for anything. They do not scratch for food or creep about their tiny, green meadows to look for grubs. When they get hungry, they merely go to a large white box from which they take whatever they want. They sit and do nothing for long periods of time, merely watching pictures on another box, which is black. They never sing, for they have smaller boxes that play music to them. Do not be surprised that their houses also look like boxes.

"The number of humans is great. I do not have the actual figure, but the turkey vulture tells me there is a great city not far from here where the number of humans is enormous. Only certain birds and mammals can live in the city—a pitiable, small number. The air always smells like smoke and oil, there is great noise at all times, and the humans, because they have nothing real to do, are unhappy.

"They wish to avoid pain at all costs and so do not exert themselves over anything. Their knowledge declines with every generation, and their ignorance is rampant. They think they are free, but they are the greatest slaves that ever were. Pleasure is their master. They imagine that so long as the impulse originates in them, it expresses their freedom. What irony. What supreme irony!

"And if you want to know what sort of beings they have become, hear my story of the two boys and the turtle."

The mourning dove drew a long and painful breath, ruffled her wings uncomfortably, and began:

"I saw it with my own eyes. One day two boys came down to Hungry Hollow from the suburb. They had come to play but found a turtle—a snapping turtle, as it happened. The turtle was crossing the meadow on its way to Hungry Creek. When it saw the boys dancing about it, the snapping turtle hissed, stopped walking, and drew its head into its shell as far as it could.

"The boys picked up a stick and began to poke the creature in the eyes. At one of these pokes, the snapping turtle seized the end of the stick quite suddenly and crushed it in its beak. This terrified the children. Then they became angry that the turtle had tried to defend itself. They went to the creek to look for stones.

"They pelted the turtle with stones, but this seemed to have no effect. How dare the turtle be immune! They went wading in the creek and succeeded in prying out a large boul-

46

der that they could barely carry between them. Then they dropped the boulder repeatedly on the turtle until its shell began to split. Blood oozed from the cracks, and the turtle made a desperate attempt to reach the creek. They drove sticks into the ground to make a pen for the turtle until they could finish their work.

"The snapping turtle, its mouth oozing blood, ceased its struggles. Its eyes grew cloudy as it neared death. The boys, hooting their victory, went home for a well-earned lunch. I flew down to the turtle and told her I had seen everything. With her dying breath the turtle said, "Please look after my babies."

"As she died, three eggs rolled out from under her tail. You can't imagine how difficult it was to . . ."

"Excuse me, Owl." It was the house sparrow. "This is really too much. Does the mourning dove actually expect us to believe that she took care of the snapping turtle's young? And how can she draw general conclusions about humanity from an isolated incident?"

"I was in the process of rolling the eggs to the creek where I could find a place to bury them. It was hard work, and the sun was going down. Just as I got to the bank, along happened Lotor. He chased me off and then ate all three eggs without so much as a 'How d'you do.' "

"Summarize," instructed the great horned owl.

"Boys will be boys. But if humans no longer grow up, can you imagine the trouble ahead?"

The Tree of Life was on the point of disappearing altogether. It shimmered and faded. The birds could barely hear each other. The owl spoke hurriedly.

"We do not have time for a vote on these charges. But we will reconvene on the next magic day."

"But wait," cried the mourning dove. "I haven't told you about the men with measuring sticks in the forest by the road. I haven't told you about the . . ."

But the Tree of Life wavered and faded and then winked out of existence. Only the hackberry was left.

Microperson

The magic day has run to afternoon. Dianne stands by the shallows of Hungry Creek, staring at the bottom, trying to visualize the life there. She wants to see it as clearly as she sees the plants and insects around her. What microscopes have revealed they have also made abstract and unreal. She tries to grasp the smallness of the amoeba as she stares into the benthos, but she cannot. She is simply too large, orders of magnitude too large.

Hungry Hollow has been having a strange effect on her lately. It has become greater than the sum of its parts, with a life of its own. It speaks to her through the catbird and the squirrel; it touches her with warm breezes that come from nowhere; it shows her new things, from the beech drops to the bagworm moth, the longer she looks. And now she slips, insensibly, into a reverie that is not her own. She wants to see the amoeba and the other lives that glide and scurry and float over the alien landscape on the bottom of Hungry Creek. If she is too large, then she must shrink.

She will do it in stages, says the magic, each stage a swift reduction in her size, a shrinkage to one-tenth in all her parts. She is on ground zero of the world she knows, the macro world. She is 1.7 meters tall: about 5 feet and 7 inches.

World One

Her eyes sting and breathing is impossible. It's her atoms. They have shrunk with her, and the molecules of air are ten times as large as they should be. She cannot breathe them. They sting the delicate tissues of lung, mouth, and eye. To avoid asphyxiation, she dons the scuba gear that appears at her feet. She straps on the tank, turns on the regulator, and tries to breathe normally through the mouthpiece. Then she puts on the face mask.

The scenery is mind-boggling. Not only is everything ten times bigger, but it's the wrong color, as though she had eaten the mushroom with the red cap. The sky is bright crimson, the huge arrowheads and bulrushes are purple, the stones green and violet, the water pink. The flowers are beacons of colored light.

The cells of her retinas have also shrunk. At one-tenth of their normal size, they no longer pick up the same wavelengths, but rather register those one-tenth as long. She now shares her spectrum with insects. Her eyes interpret shades of ultraviolet as new colors, from red to blue.

To rid herself of this unsettling landscape, she must attach a wavelength-correcting device over the front of her face mask. Now everything is back to its normal color, though very large and bewildering. Dianne is 17 centimeters tall, less than 7 inches high. She does not feel small, however; she just feels that everything else has become rather large.

She looks at Hungry Creek. It is a wide river with a fast and dangerous current. The shallows in front of her have become a bay. What appears to be a trout darting in the shallows in front of her is really a minnow. She is lost in wonderment at a landscape that resembles a tropical rainforest.

With shocking suddenness, mud now churns in the shallows-bay, and something under the water lunges at the trout. Primeval jaws close around it. Cruel eyes blink in a leathery head that swallows the fish with a single gulp. It is *Chelydra serpentina,* the snapping turtle. World One is hers. To Dianne, the monster is five meters from jaws to tail, longer than most alligators.

Serpentina lumbers off over the soft mud of the shallows toward the river-creek, raising a cloud of silt and clay. Dianne can barely see the huge clawed arms stroke as she enters the current, disappearing behind reflections of the rill-broken sky. Hungry Hollow has become a threatening place.

One thing seems familiar. There are still lots of grasshoppers around. Did they accidentally shrink with her?

They aren't grasshoppers, but springtails. (*Orchesella* is common this time of year on the bank of Hungry Creek.) One crawls slowly over the pebbles at her feet, a squat, dark insect with a long, striped abdomen. She reaches for it, but with a flick of something beneath, the springtail shoots away. How is it possible for an insect the size of a grasshopper to jump so much higher?

The answer is at hand. Since her arrival in World One, Dianne has felt unusually strong. When she picked up the scuba tank, it seemed to weigh far less than it should. The reason for this new-found strength lies in the relationships among the dimensions of things, phenomena of scale that permeate all the Worlds. It explains the enormous relative strength of insects and the immense power of swimming bacteria.

Her height shrank by a factor of 10. Her strength, which depends on area (the cross section of her muscles), shrank by a factor of 100. Her weight, which depends on volume, shrank by a factor of 1000. Thus as things now stand, she has one-hundredth the strength, but only one-thousandth the weight or mass to move about. She has become ten times stronger than she was before, relatively speaking. Once she has learned to control her new-found strength, she will be something of a dangerous character.

Dianne steps into the water. It climbs her ankles in a ring and clings to her skin. As a child, she noticed the meniscus in a drinking glass. When she grew up, the phenomenon became less noticeable. Now it is back—and it is about to get worse.

She wades into the water, drawn by the anticipation of seeing the biofilm as no living human ever has. She fears Serpentina but knows that if she becomes small enough, she will not even be a snack for the antediluvian beast. Small size is a protection. But each world has its predators.

World Two

She is now about 1.7 centimeters high, barely three-quarters of an inch tall. The water that previously came up to her ankles now immerses her chest. Worse yet, the meniscus carries the water to her shoulders—a cool, tingling cloak. It feels heavy. Water has surface tension.

The river is now very big, about 400 meters across. The waves and rills no longer look natural but seem gross and out of proportion, like soft, fast-flowing jelly. The shallows have become a small lake. Dianne ducks under, letting the surface close over her. The water has become unbelievably cold. Her body, which generates heat in proportion to its volume, has once again dropped to one-thousandth of its former mass. But it loses heat in proportion to its surface area, which has just decreased to one-hundredth of its former extent. She is therefore losing heat ten times more quickly than before. She shivers uncontrollably in a vain attempt to make up for the heat loss. The same problem plagues small mammals. The masked shrew in the floodplain meadow would die without its high metabolic rate.

She needs a wet-suit and, in her reverie, now has one. She sees massive lily pads 300 meters across the bay but does not doubt her ability to swim there. She knows they are called spatterdock, but Dianne prefers *Nuphar advena*. It sounds romantic, like an exotic destination in a travel brochure.

What a strange and magical world below! Shafts of sunlight slant through the water, illuminating a garden of giant elodea plants that look for all the world like overgrown kelp. Below her a school of young bluntnose minnows cruises at terrifying speed above the golden mud. They are the size of great white sharks and swim much faster. World Two belongs to the minnows. Dianne swims quietly.

When she comes to the lily pad, she grasps the edge to get her breath. The top of the pad has a pebbly texture, and Dianne realizes with a shock that the pebbles are individual cells. She raises her head to look at them but cannot get free of the water. It hoods her in a watery blob that refuses to fall away. A prisoner of the meniscus, she can see nothing clearly through this strange lens. She understands the struggles of insects trapped in water.

She panics when something brushes her legs. One of the bluntnose minnows has come to graze. It has rasping teeth that scrape microorganisms from surfaces. When her courage returns, she looks below the lily pad and gasps. It is not at the fearsome minnow that she gazes but at its inoffensive food. Dozens of crystal goblets hang by translucent threads from the undersurface of the lily. They are *Vorticella convallaria,* protists that she has seen only in the harsh light of her microscope. Each goblet bears a whorl of vibrating cilia around its rim. The cilia whirl like a wheel, an illusion of coordinated beating that draws even tinier organisms into a mouth in the center. When the minnow noses into their midst to feed, the vorticellas jerk quickly back, their stalks coiling instantly.

World Three

The water feels colder to her feet and hands, which are still bare. It has also begun to sting, just as the air did before. The reason is the same: molecules in motion. She needs a pair of diver's gloves now and flippers to protect her feet.

Dianne is 1.7 millimeters tall (or long), a bit over a sixteenth of an inch. The surface of the pond is still near at hand. If she could break free of its surface and look at the distant shores, she would see that Hungry Creek has become a river the size of the Lower Mississippi, four kilometers across!

Under the surface, the watery world of the "shallows" has become immensely deep. Huge elodea and coontail plants wave gently below, and now Dianne sees strange, whitish shapes swimming here and there, near and far, all through the water. Tennis balls, hotdog buns, misshapen light bulbs. Some are one-celled protists and some are many-celled animals.

Some of the protists are flagellates, motoring along with the aid of a single whip called a flagellum. Some are ciliates, driven by hundreds of much smaller whips called cilia, which coat their bodies. She recognizes *Coleps,* with grooves of latitude and longitude carved into its surface. It not only resembles a glass hand-grenade but at the moment is about the size of one.

She check her tanks (a half-hour left), expels some air, and begins her dive. The lily pad, a vast overhead shadow the size of the Columbus Circle, retreats upward as she sinks into the depths of the shallows. She is on her way to the benthos.

Down and down she drifts, so slowly that she feels suspended. The water is clear. Below her, the bottom is still 250 meters away. It approaches imperceptibly.

Here comes a beautiful translucent sphere that seems to ascend past her. About as wide as her thumb, it sports delicate glass spines radiating outward from the transparent cell wall as it drifts past. It is called *Actinophrys,* a freshwater radiolarian of sorts.

She seems to have wandered into someone's exotic offshore fishery. A huge, shrimp-like creature comes into view, swimming on its side. Being larger than Dianne, it looks formidable. But the scud or sideswimmer called *Gammarus* feeds mostly on decaying matter from plants, animals, and fungi at the bottom. It belongs to Amphipoda, a class of crustaceans.

Now you see it, now you don't. A flash of silver scales, a shape too fast to catch, and the sideswimmer has vanished. The diet of the bluntnose minnow is not confined to *Vorticella.*

The bottom is now only a hundred meters below. Dianne descends past an *Arcella,* common in the shallows. She sees its hemispherical house, a hovering UFO. The amoeboid creature within projects its being through a circular hatch in the bottom. She drifts down until she can look up from below. Clear, jelly-like arms spread out from the hole. *Arcella* is drifting—and fishing. When the arms contact something edible and not too strong, they will engulf the prey and welcome it home.

Dianne arrives with the gentlest of bumps on the bottom. The scenery is breath-taking, but she is distracted by the sight about her feet. A multitude of tiny yellow and green jewels sparkle in the sunlight, a precious gravel of diatoms.

World Four

The diatoms are much larger now, but she pauses to consider her scale before picking one up. She is barely one-fifth of a millimeter tall, a mere speck about the size of a period. How ironic that she must now measure herself in the dimension she is most familiar with. The micron is one-thousandth of a millimeter, and she is 170 microns long. The knowledge makes her feel insignificant, lonely, and somewhat trapped.

The lily pad is now 2.5 kilometers overhead! If she could stand on the pad and look around, the near shore of the shallows would be a kilometer away and the far shore of Hungry Creek some 40 kilometers away—wider than any river on Earth.

If it weren't for her wavelength goggles, her eyes would see by the dubious light of x-rays. The light is still good, and the clarity of the water reveals another colony of *Vorticella* not far off. Now they suggest transparent church bells straining upward to the height of a house, on stalks like transparent garden hoses.

She has landed on a patch of debris, nestled in silt that in World Four resembles a jumble of coarse, sharp, crystalline rocks, and boulders. Underfoot, the debris is soft and spongy, carpeted with diatoms and other algae, a brownish-green lawn that rolls darkly into the distance.

Beneath the hummocks of debris, she senses activity. No matter where she looks, there are stirrings. An indistinct form emerges briefly and then disappears again. Far from being a passive layer of steadily accumulating garbage, this landscape betrays an enormous recycling operation in which nothing, ultimately, goes to waste. As a mark of its efficiency, even the recyclers get recycled.

Bacteria and fungi invade every pore of the debris, each pursuing its specialty in the act of breaking down the bodies of formerly living things. At the same time, rotifers, ciliates, and flagellates prowl the sediments for bacteria and for each other. It is prime hunting ground, the debris. It is the place where . . .

She must stand aside quickly, for a nematode has burst from the depths right beside her. It's about ten meters long and very hungry. When its head breaks free, she has never seen anything so frightening. *Tylenchus* thrashes blindly, its three-lipped mouth opening and closing on nothing. It is a manic, translucent python. *Tylenchus* swings its head with wild vigor right and left, as though confused about the lack of debris. Then it abruptly plunges back into the landscape. The rest of its enormous, faintly mottled body snakes away below. It's a good time to take a walk. She is in the World of the nematode.

As she bounds from hummock to hummock, she frequently encounters long transparent tubes the width of a field-tile, extending for hundreds of meters. The tubes are divided into cylindrical cells. Some have spirals of green ribbon inside them, reason enough for the name that comes to her: *Spirogyra*. Other tubes have green star-bursts inside them; she knows them as *Zygnema*. She finds their precise shade of green strangely comforting in this otherworldly landscape. *Spirogyra* and *Zygnema* have the same kind of chlorophyll as the "higher" plants and may even be ancestors of them.

More impressive still are the smaller, blue-green tubes of *Oscillatoria,* animated vines in the benthic forest. *Oscillatoria* turns incessantly on its long axis, slowly drilling ahead with immense strength. When it bends upward, no amount of debris can resist its power. A huge raft of recyclables suddenly levitates in the landscape. Then she sees the blue-green pipe beneath.

Speaking of bacteria, here and there at her feet are small jellybeans, rods, and balls clinging to the rotting plant fibers and cast-off body parts of every conceivable aquatic creature, from dragonflies to rotifers.

World Five

She stands on the debris, now 17 microns tall. Far overhead, Hungry Creek is 400 kilometers across, as wide as the Caspian Sea. With each new world, she has noticed an increased sense of energy. The ratio of strength to mass has been continually rising, while the ratio of strength to drag resistance from the water has remained about the same. It takes less energy to propel an organism the smaller it gets. In World Five, some of the faster ciliates now travel at speeds of 100 kilometers an hour.

Dianne simply didn't see it coming. She has been eaten by a *Coleps,* the hand-grenade monster, now larger than she. It came out of nowhere and forced its cage-like mouth around her, pushing and sucking until she was inside. She does not panic, but she feels sorry for the creature, for she knows she must kill it. Before it whirs away with her inside, before it tries to digest her, she must find and disrupt the nucleus, a large speckled sphere nearby. She reaches through the membrane forming around her and pokes her hand into the nucleus, clenching and unclenching her fist, destroying the DNA that directs the cell's activities. The *Coleps* becomes by degrees less active. The micro-machinery grinds to a halt, and *Coleps,* turning small and abortive circles, sinks gently to rest on the debris. No single-celled creature can survive long without its nucleus.

Dianne kicks her way through the ciliate's cloying cytoplasm toward the mouth and then breaks through the tough membrane that covers it. Thus she frees herself and turns to inspect the fearsome predator. The *Coleps* is leaking badly. Its cytoplasm streams slowly into the water, a cloud of ribosomes, food particles, and the mitochondria that powered its existence.

At first, just one of the jellybean bacteria shows up. It bobbles at great speed around the leaking *Coleps*. It is an *Aeromonas,* a very flexible genus of bacteria with many trades. *Aeromonas* will eat almost any simple organic, energy-containing substance it can find. Soon other *Aeromonas* show up for the feast. They too begin to wobble around the damaged cell, seeming excited. Before long, not only *Aeromonas* is around but also *Bacillus* and other genera, including a lone *Escherichia,* the genus of bacteria that live in human intestines. The bacteria all eat invisibly and without mouths. They secrete enzymes that break down the carbohydrates leaking from the dying *Coleps.* The bacteria then absorb the simple sugars that the enzymes create.

On an impulse, Dianne seizes an *Aeromonas* as it motors by. It's quite a handful. Shaped like a curved party balloon, it sports a flagellum at one end that whirs like a weed-eater. Inside this bacterium, within the inner wall, runs a small, circular, biochemical motor powered by ATP.

Before she lets the bacterium go, she looks through the clear cell wall. She sees no nucleus, but she did not expect to. Bacterial cells have no nuclei. They lack the husk, or karyos, that encloses the DNA in the cells of eucaryotes: fungi, plants, animals, and protists. The chromosome of *Aeromonas,* like that of all procaryotes, is a continuous loop that winds throughout the cell. Although she sees only little specks and bubbles inside the bacterium, she knows them as the superstructures of an entire chemical factory.

Off goes the *Aeromonas*. Dianne has tripped on what appears to be a soft, coiled pipe, pale blue-green in color. It is *Lyngbia,* a cyanobacterium. She notices a tiny icosahedron stuck to the bacterial wall and nearly swallows her regulator. A virus! She longs to visit the next world, but she is almost out of air. It is just as well that she leaves, anyway. In World Six she would be pounded by individual water molecules. For protection, a suit of medieval armor might just work.

She drifts upward now, enlarging as she goes, breaking out of the shallows and dripping with dreamwater. She walks ashore to rejoin the material body that stands there, lost in thought.

Water

The day of magic draws to a close. The sun paints the clouds rose and magenta, then silently sets. The air cools, wringing out the water as dew. *Argiope aurantia,* the black-and-yellow argiope, sits on the zig-zag of silk in the center of her web and shakes the lines. They feel heavy. Argiope grows sluggish in the cool air. Dew condenses on her black hair and glistens on her gold bosses. She will catch nothing until morning. She grows torpid and sleeps the sleep of spiders, invisible in her bush of grey dogwood.

Water is the last act in this magic play, its secrets given away by molecular machinery behind the scenes. The first secret concerns a web much finer than *Argiope*'s. Without this web there would be no liquid water.

Without water, Hungry Creek would be a forlorn depression lined with stones, sand, and baked clay, winding through the hollow. Without water, the soil of the meadow and forest would be mud-cracked desert. Without water, the living things of Hungry Hollow would be shrunken mummies, freeze-dried fossils immobile, dead.

For those who reflect on the environment, there is no better mirror than water, no better window than ice. The world's most common substance conceals uncommon structure. Water lives far away in World Eight. Were it not for a certain angle in its molecules, all the water of Hungry Hollow would evaporate in a matter of minutes or hours.

Water molecules are very, very small. The droplet on the spider's web contains about three and a half billion of them. Only a microtourist of World Eight would witness water as it really is.

Every molecule of water has the structure called H_2O: Two atoms of hydrogen (H) embrace one atom of oxygen (O). Because the oxygen atom has twice the diameter of the clinging hydrogens, the surface of the molecule is mostly oxygen—a rounded sphere that sports two hemispherical bumps, the hydrogens. Shared electrons make up the "surface" as they zip around the three atoms, weaving a quantum web of electric charge, a fuzzy force field that resists poking. The bonds of shared electrons keep the oxygen wedded, if not welded, to the hydrogens. The marriage will last until some chemical encounter breaks the bonds.

The secrets of water lie in the shape of its molecules. The two hydrogen bumps are not attached on opposite sides of the oxygen, 180 degrees apart. Instead, they make an angle of 104 degrees. The water molecule is lopsided, and this gives it an electric polarity. There is a slight negative charge on the oxygen side of the molecule, because the electrons linger in their orbits near the positive nucleus of the oxygen atom. The molecule's hydrogen side has a balancing positive charge. The importance of this polarity, not to mention the 104 degree angle of the twin hydrogens, is hardly obvious. But it is crucial.

Like most molecules at ordinary temperatures, the triplets of water spend their lives in a state of continual motion, the expression of heat itself. In the dewdrop, in the limestone fissure, in Hungry Creek, and in Lotor's bloodstream the molecules jostle and bang. They vibrate. They slip and slide over and around each other in a frictionless dance that, left to itself, would never run down. Here is the true *élan vital,* the force that makes life possible. Without the mad dance of heat, chemical reactions would be rare events and biochemistry would be impossible.

When two water molecules meet, the negative side of one molecule seeks the positive side of the other—an extramarital longing. This attraction, which is called a hydrogen bond, amounts to a brief holding of hands, a momentary infidelity, as each molecule moves among the others. But within the water as a whole, it becomes a cohesive do-si-do of billions.

The first secret has finally emerged: Without hydrogen bonds the molecules of water would part company. Its sociability suddenly gone, the droplet of dew would evaporate in a second. Hungry Creek itself would vanish in a few hours.

The dance of the water molecules is not entirely a random affair. There are structures even in the droplet of dew. Hydrogen bonds draw the surface of the droplet into a web of molecules, each one linked to its neighbors. The web encloses the droplet in a bag so tight that the balance of forces guarantees a sphere.

The web appears wherever water meets air. It supports pond-skaters and whirligig beetles. It makes a meniscus where water meets land, the tiny liquid moulding that climbs a millimeter or two up the arrowhead stem. The web also works within the soil of the meadow as capillary action. The meniscus climbs every pore and fissure of soil, drawing groundwater toward the surface. The life of the plants and trees of Hungry Hollow depends utterly on water's web, on the hydrogen bond.

Inside the dewdrop—inside *any* liquid water—the hydrogen bonds make other structures. Chains of molecules build into conga lines. Sheets of them disco through the teeming billions. But the surrounding dance is too violent. It knocks holes in the structures and breaks them into pieces that must begin again.

In a few months, winter will come to Hungry Hollow. The temperature will drop, and energy will drain from the water of Hungry Creek. It will happen first in the shallows. The water molecules will slow down as though tiring. The chains, sheets, and other structures will grow larger and more elaborate. The hydrogen bonds will take over. At 0 degrees Celsius, the molecules will interlock in a secret lattice and the dance will be stilled.

In ice, water molecules form an intricate crystalline structure. Each molecule holds hands with four others, above, below and to the sides. The hydrogen bonds formalize into a three-dimensional lattice. Seen from above or from the side, they appear as nothing but fairy rings of six molecules. Each ring has a curious zig-zag geometry, bending up and down, alternating, three times. In the mathemagic of these crystals, each molecule is, in turn, part of six rings.

Seen from above, the rings reveal the next secret. They make perfect hexagons. If the hexagons were flat, ice might split like mica. But the bends interlock to form a powerful mesh. Ice is hard. Made cold enough, it will ring like a bell.

In the snowclouds high above Hungry Hollow, water will crystallize inside supercooled droplets. As each tiny crystal outgrows its droplet, the hexagons in one set of planes predominate. Carried by air currents along a unique path within the cloud, the crystal will add water molecules at all points of growth. In this way it builds outward to visibility, becoming heavy. It begins to fall, uniquely engraved with its history, writ in hexagonal script.

The ice sheet that will cover the shallows hides the last secret. When the lattice of interlocked hexagons first forms in water, the angle between the hydrogens in each molecule will yawn ever so slightly before the sleep of winter. Half of the hydrogens will pull somewhat away from their oxygen partners, reaching out to other oxygens. In the new geometry of interlocking rings, the angles of individual molecules of water have changed from 104 to 109 degrees—small cause for such a big effect. The entire mass of water will expand as it crystallizes. In a crack in the limestone ledge, the capillary water will freeze and shoulder into the rock, strong enough to snap a piece of it off.

The water that expands into ice will have the same number of molecules but a larger volume. It will float in the shallows. Most other liquids produce ice that sinks. If the water of Hungry Hollow sank, it would coat the bottom first and then gradually build to the surface, choking the flow until the water spilled over the banks to freeze in the meadow. Fish, crustaceans, and most of the aquatic insects would die. The absence of such disasters is reason enough to be grateful for a small change in a tiny angle.

It is night. The water of Hungry Creek appears black. Now is just as good a time to examine its molecules, for in World Eight there is no light, as such. Every photon carries false information, altering the thing it reflects. There is only the light of imagination, and by it we witness a confusing array of molecules. The creek water is a chemical soup.

The list of ingredients sounds alarming: humic acid (from forest soil run-off), silicic acid (from clay and weathering feldspar), metal ions (from rocks and decaying tissues), nitrates (from the soil and from fertilizer), carbonates (from dissolving limestone, bone, and shell), phosphates (from agriculture and decaying tissue), carbon dioxide, oxygen, nitrogen, and (in amounts too small to measure) enough chemicals—some natural, some human-made—to fill a small book.

There is a reaction in Hungry Creek that symbolizes all reactions. Amid the boisterous myriads of angled water molecules, strait-laced triplets of CO_2, or carbon dioxide, make their way. Two large hemispheres of oxygen attach to opposite sides of a smaller carbon, making three atoms in a row, almost a dumbbell. No hydrogen bonding here, at least nothing approved of by the molecules of CO_2. Yet they will frequently break apart, reacting with water to make carbonic acid.

$$CO_2 + H_2O \leftrightarrow H_2CO_3$$

The carbonic acid molecule is not confined to the line its formula suggests but rather has a triradiate form. After the reaction, the carbon sports three oxygens like three blunt spokes in a wheel. Two of the oxygens have hydrogen bumps opposite the carbon. The third oxygen has no hydrogen. It shares all its electrons with the carbon, doubly bonded and snugging closer to it. Carbonic acid makes soda-pop, but it is unstable. It may prefer to be carbon dioxide again, which is why soda-pop fizzes.

The reaction goes both ways. At all times, there are carbon dioxide and water molecules making carbonic acid. And at all times, carbonic acid molecules are turning back into carbon dioxide and water. But the reaction proceeds more frequently in the former direction: a lopsided equilibrium. At any moment, the number of carbonic acid molecules is somewhat larger than the number of carbon dioxide molecules, and the ratio of the two species always hovers around the same value.

The disintegration of carbonic acid back into CO_2 and water is a downhill reaction. It gives up energy. A molecule of carbonic acid, H_2CO_3, is greater than the sum of its parts. It has slightly more energy in its five bonds than a molecule of water and a molecule of carbon dioxide have in their four bonds combined. Given a choice, nature always runs downhill. Only the supply of heat energy in Hungry Creek keeps the supply of carbonic acid plentiful.

"Downhill" and "uphill" are keys to the processes of life itself. In the darkness of the creek there are organisms large and small, their biochemistry running slowly downhill like so many wind-up toys. In the morning, the sun will begin to wind the green ones up again.

At last the sky in the east turns orange, then yellow. The first warmth touches the droplets on *Argiope's* web. In each droplet, a much smaller web of water holds fast. Molecules of water from the humid morning air zip into the bag from the surrounding aerial dance. Through this process of condensation, they take a position in the surface web, identical to their new partners. At the same time, molecules inside the droplet that sport an excess of energy break free and join the surrounding humidity; they evaporate. There has been an equilibrium through the night. But as the droplet heats up in the morning sun, the equilibrium shifts. More molecules leave the droplet than enter it. It shrinks imperceptibly until it vanishes. *Argiope* shakes the web and finds it light. She is ready.

Cymbella and the Hypotrich

The bed of Hungry Creek is a vast, liquid world of fields and forests, deserts and wastelands over which the water flows like an unending hurricane. Closer to the bottom, it moves in ever-slower boundary layers until it becomes an aquatic breeze to the billions that live in the biofilm.

Between two rocks in the center of the stream there is an especially pretty stretch of country that may as well be called the Serengeti Plain. In the measure of World Five, it is a wide, algal savanna some two kilometers across. It is beautiful, like its African counterpart, but far more violent. Though it all seems peaceful enough from a distance, in this sparkling, sunlit, olive-green world are committed acts of savagery and appetite that dwarf anything in the familiar dimensions of World Zero where humans live. It is here, in these sunny fields, that you will find *Cymbella* and the hypotrich.

Cymbella ventricosa is a diatom. Officially, she is not a "plant," but in her fundamental role of producing energy-filled carbohydrates from sunlight, she works just like one. Hers is but one of 67 species of diatoms in the Serengeti at the moment, and she is herself but one of a hundred thousand individual diatoms living there. Their glass walls sparkle in the sunlight; their lime and gold pigments blend subtly with the darker greens of the place. They are not rooted but trundle over the soft and uncertain ground amid the blue-green cyanobacteria that wave in the currents like wind-blown grass.

The hypotrich is called *Oxytricha,* a multilegged protozoan beast that roams the Serengeti in search of prey. He is not officially an "animal," but in his fundamental function, converting carbohydrates to energy, he works just like one. His species is one of seven hypotrichs in the Serengetti, and he is one of a thousand individuals who ply the same general trade. They prowl through the day and night, searching, searching for something to eat. They run on transparent, flexible legs that twitch and flick in frightening fashion. There is no nervous system operating the legs or any other part of the hypotrich. Nothing to be seen on other planets could appear half so alien.

If he eats well for a lifetime (a day or two), *Oxytricha* will divide in two. A pinch across his middle will appear, and within half an hour, there will be two hypotrichs where before there was one.

Cymbella and the hypotrich subscribe to an ancient pact that neither can dissolve. Each makes what the other needs, and thus they embody two fundamental processes: photosynthesis and respiration. These are the main engines of life, not just for *Cymbella* and the hypotrich, but for all organisms on Earth (except some bacteria). *Cymbella* makes oxygen not only for *Oxytricha* but for all respirers, including even herself. And *Oxytricha* respires not

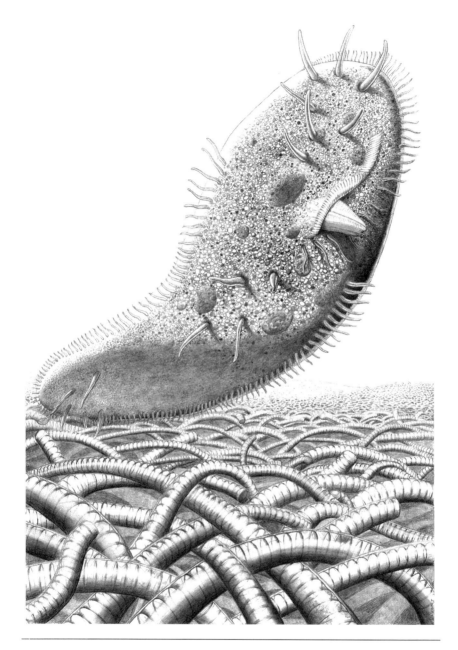

"Cymbella *and the hypotrich subscribe to an ancient pact . . .*"

only the oxygen released by *Cymbella* but that produced by all pigmented producers, from diatoms to plants. Wherever it came from and wherever it's going, oxygen is oxygen.

Cymbella consumes carbon dioxide and water to produce sugar and oxygen. *Oxytricha* consumes sugar and oxygen to produce carbon dioxide and water. What one produces, the other consumes. The carbon, oxygen, and hydrogen of these compounds cycle endlessly between the two kinds of organisms, producer and consumer. But the essence of the pact is one-way. It is not about atoms or molecules, after all. It is about energy.

Cymbella positions herself in a sunny spot, absorbing light in her chloroplasts. There, molecular antennas capture photons from the sun. Only photons of certain wavelengths are caught, the rest rejected. Green chlorophyll absorbs red and blue wavelengths, reflecting the unwanted color. Although the green is rejected by the forest, it finds a home in our eyes. Gold and lime paint the fields of diatoms, for these are the unwanted wavelengths of *Cymbella*'s chlorophylls.

Each photon that *Cymbella* absorbs displaces an electron that powers a complex cycle of reactions. Like a hot potato, the electron is passed from one specialized molecule to another, splitting water into hydrogen and oxygen in the process. It returns, ultimately, to the chlorophyll molecule. In the meantime, it has lost energy to chemical reactions that succeed in producing a molecule of ATP, a packet of chemical energy that powers a myriad of reactions. *Cymbella* puts most of her ATP to work assembling molecules of glucose from the carbon dioxide she has absorbed from Hungry Creek. The carbon, oxygen, and hydrogen are arranged into six-sided rings of glucose. Its bonds hold the energy she has captured from the sun.

$C_6H_{12}O_6$

She assembles the molecules of glucose into rafts of a carbohydrate compound called chrysolaminarin. *Cymbella* will store the glucose in this form until she (or someone else) needs it.

In the process called respiration, *Oxytricha* must consume and digest carbohydrates (among other things). He has enzymes that break down the carbohydrates to produce glucose. A complex chemical cycle then takes over, converting the glucose to simpler compounds in a downhill reaction that releases the bond energy of the glucose into a new molecule of ATP (adenosine triphosphate), the chief energy carrier of all cells. With this energy, *Oxytricha* scurries forward on transparent legs to hunt. Indirectly, then, he is solar-powered after all.

Cymbella

If you look more closely at the benthic meadow, you will see not only a delicious riot of green, gold, and every shade in between, but also an endless litter of glass shapes glittering in the sun: violin cases, sarcophagi, candy dishes, canoes, and pill boxes. They are not passive as the plants of our macro-world, but occasionally glide, self-contained, seeking that optimum patch of sun or shade.

Cymbella is assymetrical, pointed at both ends, but slightly curved, like a banana. She lives, as other diatoms do, between two transparent lids called valves that fit, one inside the other, in the manner of a glass jewel box. One lid is larger than the other, and this produces a peculiar effect. When *Cymbella* reproduces, the protoplasm within her divides. The two valves then come apart, and each of the daughter cells attaches to one of the valves, growing a new valve to fit inside the one she inherited. The daughter who got the larger lid will fit a new one that duplicates her mother's lesser lid. But the daughter who inherits the smaller lid must make an even smaller one to fit inside, so she is smaller than her sister.

Reproduction by simple division is called vegetative growth. In diatoms this mode of multiplication leads, generation after generation, to individuals that become smaller and smaller, on average. At a certain point, the distant descendants, cramped and small, must give up the shell game for sex. The smallest *Cymbella* shed their valves, transforming themselves by division into mobile organisms called gametes. The sexual cells have inherited only one set of chromosomes, and when they fuse in pairs a new organism results: a would-be diatom with the normal, paired chromosomes. The new diatom enlarges as much as it can, taking in extra water, in order to build new, full-size valves. It must do this as quickly as possible, before something eats it.

Cymbella makes new valves from silicon in the water around her. At her naked membrane, *Cymbella* adsorbs molecules of silicic acid, a component of clay. She re-forms them into a delicate blend of silica and water that builds to a tough, aqueous glass decorated with hundreds of tiny holes, a latticework of ancient design through which she can pass molecules in and out. In World Five, the valves are proof against most marauders.

Beneath the glass, *Cymbella*'s cytoplasm restlessly churns. Directed by her unicellular brain, the nucleus, the cytoplasm keeps the green and golden chloroplasts supplied with carbon dioxide and water. It respires, as all living tissue must, but has the luxury of its own carbohydrate supply. It manages motion, too. The lower valve has raphes, two long slits that run from either end to her middle. Who will finally fathom how diatoms move? Is Cymbella high-tech? Does she use electrostatic propulsion? Or does she walk on temporary fingers, little filaments protruding through the raphe? However she does it, you will see her in the meadow, trembling now and then to a better position, her emeralds and gold on display: chloroplasts, carbohydrates, and oils precious beyond reckoning.

Oxytricha

Like *Cymbella, Oxytricha* is a single cell. An advanced ciliate, flat and sleek, he legs over the microsavanna on cirri, fused cilia that resemble slender, flexible triangles. The tips engage diatoms and detritus; the bases flex with many microtubules operated by a system of protein rods just beneath the cell wall. *Oxytricha* is long, with a rounded front and rear, a transparent sun-racer.

He rushes forward on three sets of cirri: front, middle, and rear. He moves with incredible speed, stops on a dime, and then reverses himself. If he draws near, you will see a strange beating motion instead of a front bumper. A row of cirri follow the graceful curve around one side, then under. The cirri end at a slit-like mouth constantly opening and closing beneath the creature. *Oxytricha* feeds on anything it can ingest and digest: particles of cellular debris, small diatoms and algae, other ciliates, amoebae, flagellates. Lion, jackal, and antelope in one, it attacks, it scavenges, it grazes. It is voracious.

The internal anatomy of *Oxytricha* is all visible within. The nuclear area, seat of the genome and metabolic instruction, is divided into two bean-shaped compartments called macronuclei. On one side of his body, a bubble swells slowly inside the creature's wall, distending it. Then the bubble suddenly collapses, jetting waste water into the aquatic breeze. Protist kidneys! Scattered throughout his interior are food vacuoles. Spheres that contain remnant lunches. Here are some barely recognizable fla-gellates and some half-digested remains of *Oscillatoria*.

The Rape of *Cymbella*

It might be said that *Cymbella* barely noticed when *Oxytricha* scurried over her upper valve and paused to feel the diatom be-neath him. Being ten times as long, he felt her to be just bite-size and positioned himself so that his oral cirri coaxed her into his buccal cavity. Being a smallish diatom, she slid in fairly easily, and *Oxytricha* rushed toward his next victim, wherever that might be.

Carried off this way, she sensed little motion and knew only the luxury of uninterrupted sunlight. There was no need to move one way or the other to avoid obscuring organisms or debris. As Oxytricha formed a membrane about his prize, Cym-bella relaxed, kept her photosynthetic factories going, and pro-duced oxygen. Thus supplied internally, *Oxytricha* could in-crease his respiratory rate even further. For the better part of the afternoon (a lifetime for many protists) *Oxytricha* and *Cymbella* enjoyed the cooperative part of the equation. He got her oxygen and she got his carbon dioxide.

Their relationship was not platonic but intimate and symbi-otic, at least for a while. In this, they re-enacted a crucial step in the evolution of life. It is thought that primitive bacterial cells, prokaryotes, ingested other primitive bacterial cells to make the first eukaryotes. Among the ingested cells were pho-tosynthetic bacteria and sugar-metabolizing bacteria. The for-mer live on today as chloroplasts in higher algae and all plants. The latter have become mitochondria, the energy-producing or-gans of every eukaryotic cell from *Cymbella* to Dianne.

By the end of the day, the relationship between *Cymbella* and the hypotrich underwent a sea-change. As a result of swallowing so many oxygen bottles, *Oxytricha* needed carbohydrates as never before. Digestive enzymes, delivered in little packets from neighboring ribosomes, entered Cymbella's compartment, infiltrating the latticework of her valves, and setting to work on the chrysolaminarin reservoir. The carbohydrates were snipped back into molecules of glucose by a million chemical scissors. Well into sunset, *Oxytricha* digested not only her carbohydrates but also her proteins and nucleic acids, fatty oils, and everything else, until only her silicate valves were left. Just after sunset, he ejected the plundered jewel box from his mouth and scurried away over the darkening grass of *Oscillatoria*.

The rising moon threw a fitful and wobbling light on the Serengeti. *Oxytricha* would live and reproduce again, extending by one generation a line that began long before the Cambrian Age. *Cymbella* was not entirely gone, however. She had a thousand sisters.

On the Back
of a Turtle

In the Hindu religion, the world rests on the back of a great turtle. In native American religion, the world is called "turtle island" for the same reason. In either case, we humans are much smaller than the great turtle. A look through the other end of the telescope produces a normal-sized turtle with truly tiny creatures inhabiting its back. Such is *Chelydra serpentina,* the common snapping turtle.

She lives on the bottom of Hungry Creek. By day Serpentina may lie quietly in midstream where she is easily mistaken for a large, mossy rock. Or she may bury herself in the muddy shallows, a flat stone with a bit of wood sticking out.

Above or below the water, she watches for things that move, being expert in preying on a dozen phyla. She knows them not by name but by taste: tadpoles and pollywogs, minnows, madtoms, crayfish, earthworms, snails, and clams. She will take, when offered, baby ducks and grebes and, every now and then, the paw of a raccoon.

Lotor is wary when he fishes and does not like to wade too far out, for Serpentina roams the creekbed by night, shedding her disguise and abandoning her sedentary ways. She lumbers slowly across the rocky, moonlit bottom, plying her deadly trade with an oversize head that can dart forward at lightning speed and jaws that can crush a broomstick to splinters. At her full adult weight of 60 pounds, she will be nearly the alligator of Hungry Creek.

For vitamins and additional carbohydrates, Serpentina frequently adds a dainty salad of water plants or algae.

Her daytime disguise as a large, flat rock depends on the world she carries with her. Each of the hexagonal plates that compose her carapace resembles a modest mountain, rising from a plain to a sharp peak in the middle. Made from bone and covered by a tough outer skin, these mountainous tracts support an enormous population of living forms, interlaced kingdoms: Monera, Protista, Fungi, and Animalia. Serpentina's back resembles a rock because it supports almost the same community as the rocks around her. Consequently, she has just the right mossy, hoary look.

As in a vast forest tract, plants (of a sort) are everywhere. The moneran plants all belong to one phylum called Cyanobacteria. Chief among these are species of *Oscillatoria* and *Spirulina,* blue-green vines of the watery forest. Each strand of *Oscillatoria* is made of hundreds of coin-shaped cells stacked together in a long, smooth cylinder. Each compartment is a cell that captures sunlight, making starches and sugars as plants do. However, cyanobacteria use a different type of chlorophyll, an eerie blue-green compound spread evenly throughout the cell and not gathered into chloroplasts.

The oscillatorian cylinders glide slowly, gently rotating as they go. But *Spirulina* is a spiral and corkscrews ruthlessly through all obstacles in its continual search for sunlight. Always seeking the brightness that means life, the cyanobacteria weave themselves into a mat, a writhing blue-green meadow in which their moneran cousins, the eubacteria, may pursue the fine art of decomposition. The cyanobacteria dwarf the eubacteria, but the latter make a subkingdom of their own, and many of its phyla are represented on Serpentina's back.

Other, more familiar greens in the forest come from the phylum Chlorophyta, algae that use almost the same photosynthetic process as the flowering plants of Hungry Hollow. *Basicladia chelonum* is one of these and is the main tree of Serpentina's forests. Attached by a holdfast to a bit of protein ground, *Basicladia* makes branches that wave in the microcurrents. It has a close relative called *Cladophora* that clads the distant worlds on streambed rocks, but *Basicladia* is peculiar to the backs of turtles.

As Serpentina snoozes in the shallows on a sunny August afternoon, the light shoots through the forest on her back, burnishing the *Basicladia* trees and illuminating the blue-green vines of *Oscillatoria* and *Spirulina* that twine among their trunks.

Scattered throughout the forest floor are the diatoms such as *Cymbella*. They are gentle, bumbling things, no threat to anyone. Yet some creatures threaten them. Besides the nimble *Oxytricha,* there is *Frontonia,* a large, flattish relative of *Paramecium,* as well as *Actinobolina* and *Spirostomum,* the anaconda of the ciliates. The phylum Ciliophora is the most diverse of all the consumer phyla that inhabit the turtle's back (as well as the creek itself). Ciliates are generally larger than diatoms, but like diatoms, they tend to be transparent. In a world of no vision they have little to hide. Their shapes are more various, and words do them less justice: To say that there are torpedos, barrels, globes, shoes and slippers, ribbons, and cones hardly begins to describe them. All members of the phylum Ciliophora have numerous swimming organelles called cilia. Tiny whips, arranged in rows, patches, or circles on the outer wall, beat rapidly and with startling coordination. Ciliates swim with an impressive smoothness and rapidity.

Coleps hirtus, the hand grenade ciliate, has a rigid outer skin, or pellicle, that is corrugated in two directions, girding the creature in a grid of squarish, raised plates. The cilia beat between the plates in rows from front to back. The front of *Coleps* is ringed by staves, extensions of the frontal circle of plates. In the center, between the staves, is the creature's mouth.

Coleps swims between the trees with frightening speed. It arrives above an *Aspidisca,* a smallish hypotrich (also a ciliate) that has paused to rest. It moves upon the quarry, chemically sensed, before the *Aspidisca* can jerk backward in the manner of hypotrichs. It catches the hypotrich in its staves and applies a sucking mouth. The hypotrich then bends and breaks, even as the *Coleps* distends. Before another minute passes, *Aspidisca* is half inside its predator.

But *Aspidisca* has leaked a little, and the chemical message attracts other *Coleps* to the kill. Three more swarm in to seize what is left, and in the ensuing tug of war, the four *Coleps* whirl and jerk. The feeding frenzy puts sharks to shame. In a few more seconds, *Aspidisca* is but a memory.

The cilia that propel these protists, drawing them ever forward, have another use. If a ciliate anchors its rear, the beating of the whips will create a current that brings any and all floating foods to the grooves and pockets that pass for mouths. The sessile option has been chosen by a ciliate called *Epistylus chrysimidis*. It also lives mainly on the backs of turtles and rarely elsewhere. It is enormous, relatively speaking.

Anchored by a transparent tube to the turtle's back, it resembles a light bulb with a spiral groove that encircles it. As the groove nears the upper end, it becomes a helical staircase of death. Cilia line the rim of the groove and beat incessantly. The incoming current carries potential food down into a large gullet where the cellular mind must decide what is good and what is not.

From *Epistylis*'s anchoring tube, other epistylids also sprout. Whole trees of *Epistylis* are a common sight on the turtle's back. No trees in the Hungry Hollow forest, it will be noted, sprout meter-long light bulbs that suck in passing birds. *Epistylis* has a close relative on the rocks of Hungry Creek. *Vorticella,* the bell-shaped sessile feeder, was seen by Dianne one magic afternoon.

The kingdoms of Monera and Protista do not quite account for all of Serpentina's microcosm. There are true animals here, as well, but animals like nothing on macro-Earth. The rotifers, which exhibit the same range of sizes as ciliates, consist not of one cell but of many. At first sight they resemble citizens of the simpler kingdom, being transparent and feeding with cilia. But inner compartments for digestion and reproduction tell a more complex story. They are multicellular. Their cells are so small, however, that large tracts of the rotiferan anatomy have no cell walls and exist merely as a tissue of cytoplasm and scattered nuclei.

The rotifer *Trichocera* has a shell like Serpentina's. A transparent casing called a lorica protects everything but the protruding wheels of cilia at her front and a long foot that extends from the rear. Like most rotifers, *Trichocera* can attach herself to anything by extruding a fast-setting glue from the end of her foot. The wheel at the front begins to turn, and from the flux *Trichocera* selects organic debris, bacteria, and small pro-tists, swallowing them. She resembles a floor polisher with an attitude. Down her gullet go debris and creatures, grist for a fearsome mill called the mastax. Two jaws gnash incessantly in a chamber inside her, grinding and chopping everything into bits too small to see even in a microscope.

When food grows short, the time has come for *Trichocera* to leave. She dissolves the glue that holds her down, and the beating wheel becomes a propeller that draws her into new ad-ventures among the branches of *Basicladia* and *Epistylis*.

Lest we imagine the animals to be dominant here in World Six, let us watch the ingestion of a healthy rotifer by *Actinobolina vorax,* a bumptious protistan bully that stings its prey with filamentous toxic tentacles, immobilizing it. The frozen *Trichocera* is twice the size of *Actinobolina,* but no mat-ter. Much genetic "intelligence" is evident in a show that lasts for fifteen minutes as *Actinobolina* struggles to ingest the ro-tifer, finally pinning it against some debris and pushing, push-ing, until the mouth dilates to slip slowly around the sharp foot and humped lorica. It is like fitting a garbage bag around a motorcycle.

There are many other phyla, especially of plant-like organisms, hanging for structural support on everything. The trees and vines are decorated with star-shaped *Pediastrum,* spined *Scenedesmus,* and beautiful crescent *Closterium.* A colony of *Dinobryon,* a detached tree made of wine glasses, swims blindly into view. Apple green colors of the flagellates that dwell in each wine glass speak of yet another mix of magic pigments to gather sunlight.

The long and the short of life on a turtle's back, like life everywhere, lie in the production of food by everything that has pigment: green, yellow, blue-green, gold, apple green, or even red. The consumers lead active lives, eating first the producers and then each other. Nor is there peace in death, for many organisms subsist on the dead. There are protistan fungi in the phylum Oomycota and multitudes of bacteria that can break down virtually every cellular material. To be absorbed by bacteria or fungi is not an ignoble thing but a gentle dissolution into chemical constituents. Recycling is the afterlife.

In all this production, consumption, and recycling, the most general theme seems to be delay: Keep the energy of those chemical bonds from dissipating as long as possible. Keep the ball bouncing. Some atoms may recycle thousands of times before finally emerging from the community aboard the turtle in expelled carbon dioxide, oxygen, or nitrogenous compounds. The community cooperates unconsciously in this grand game. Energy is precious. There is nothing else.

The community also cooperates unconsciously in the pact the great turtle has made with them. "Make me look like a rock and I will keep you in sunlight."

The Meadow

Serpentina would go a-wandering. She emerges from the shallows on a hot, dry afternoon. The world on her back enters a crisis. A watery film compresses the forests and fields. There is barely room to swim. The protists and the rotifers build cysts, hard capsules in which they will curl up with what little water they can include. The algae prepare to dry. There is little time left.

She crosses the bar of the shallows and pauses under a thicket of grey dogwood to rest for a while on a patch of delicate moss. Stopping frequently is a good idea. It gives her time to consider where she will go next, and it gives small animals who do not recognize her an opportunity to wander past. As she waits, her shell and skin dry out. She takes on the hue of cured cement and comes to resemble a rock. But her profile, hooked beak, and tiny eyes are pure reptile. Serpentina deserves our respect. She is a living fossil, older perhaps than the dinosaurs themselves. Ancestors remarkably like her waited calmly in pools while *Tyrannosaurus rex* thundered overhead.

They survived the bitter cold following the great asteroid collision that ended the Cretaceous period.

The world on her back has gone into seclusion, hiding in cysts, temporarily dry, or dead. But a new world may be found under her feet. The bed of moss is a forest of startling beauty in World Four. Great, glistening leaves one cell thick mount tall stems in whorls and rows. Giant mites and millipedes stalk the bright, glistening foliage. And in every meniscus pool at the bases of the great moss-trees, rotifers and protists swim like ciliated fish, hunting bacteria and each other.

The moss of this particular forest, *Fissidens viridulus,* is a member of an ancient phylum that has persisted even longer than the reptiles. The bryophytes are rooted in the Devonian world and probably earlier. Before the Acadian mountains rose to the east of Hungry Hollow, the mosses and liverworts had already invaded the land, their constant need for water betraying an aquatic origin. *Fissidens* recapitulated this invasion when it grew last spring. It began as an inconspicuous, branching filament closely resembling a green alga. From this filament small shoots grew upward, much larger than the filament and several cells thick. The shoots gave off leaves, but they had little in common with the leaves of larger plants, for they had no veins. Fluids do not need mass transit in such a small plant; they simply diffuse from cell to cell.

There is a secret buried in every cell of *Fissidens*. Like all bryophytes, this most visible form of the plant is haploid. Every nucleus of every cell carries just one set of chromosomes instead of two. Because it is destined to produce male and female gametes, or sex cells, the plant is called a gametophyte, or gamete-plant. The top of every shoot produces either a complex of microscopic clubs called antheridia, which make sperm, or a tiny vase called an archegonium, which makes eggs. When the weather is particularly wet and development of the reproductive structures is complete, sperm may swim from antheridia to archegonia, entering the necks of the vases and fertilizing the eggs.

There follows an abrupt shift in gears, genetically speaking. The fused sperm and egg, or zygote, now divides repeatedly, every new cell containing not one but two sets of chromosomes. The new, diploid organism that develops belongs to a fundamentally different generation. It is the sporophyte—a separate plant, a parasite of sorts. The sporophyte takes root in the substrate of the old vase as though it were so much soil and grows a single stalk upward, far above the diminutive shoot. A capsule develops at the end of the stalk, and inside, thousands of spores mature. The spores are all haploid, seeds of a new gametophyte generation. When a lid on the capsule opens and the spores are liberated to the winds, the sporophyte's job is done. Its brief time in the sun is over.

Serpentina decides the pool must lie straight ahead. She plods over the *Fissidens* moss, a glistening bed of gametophytes with two or three sporophyte capsules sprouting here and there. She crosses some grass, paralleling the bank, and then passes a small stand of horsetail plants, also relics of an ancient development in the plant kingdom. They belong to the phylum Sphenophyta and reproduce rather like the plants at Serpentina's next stop. Serpentina continues until she rests, nearly lost, amid the giant fronds of the royal fern, *Osmunda regalis*. Here is scale restored: A reptile among the filicinophytes was a common scene in the Triassic period.

The phylum Filicinophyta arose in the Devonian period after Bryophyta was off and running. Here the gametophyte and sporophyte generations also alternate, but a shift has occurred. The gametophyte is no longer the dominant form. Search carefully and you might find the inconspicuous, heart-shaped thallus of the royal fern gametophyte. It hugs the earth, sending tiny rootlets into the soil and producing, in time, several bumps on its green surface. Closest to the tip of the heart, antheridia produce swimming sperm. By the cusp of the heart, a few archegonia swell into cloistered chambers, each for a single egg. Fertilization of the eggs of *Osmunda* requires some wetness, just as it does for *Fissidens*. Because sperm must swim to eggs, ferns also prefer damp places.

The fertilized egg, or zygote, now becomes the new sporophyte generation, a mass of cells that swells into a fiddlehead, uncurling its fronds into a new world of sun and shade, as ferns have done for four hundred million years. If size means anything, the sporophyte triumph is nearly complete. The fronds of *Osmunda* may reach two meters or more, evoking the Devonian forest.

Serpentina watches a very foolish leopard frog with her beady eyes. It simply fails to see her, as amphibians have been doing since turtles first evolved. Frogs, themselves predators of insects, need motion to distinguish objects properly. Serpentina doesn't move. She is a rock. But suddenly, part of the rock becomes a blur that the green frog's brain simply cannot process quickly enough. It does not jump in time. Two legs struggle briefly from a corner of Serpentina's crushing beak, a ridge of very hard bone that cuts anything but rock itself. Balancing the stunned frog on her tongue, she thrusts her head forward again, around it. The frog is gone in a swallow.

Her wandering has been a good idea, and Serpentina would like to wander some more. The pool is surely ahead but not near the shore. She turns toward the meadow, out of the shade of the ferns, the dogwoods, and the willows.

The sun beats down. Serpentina's temperature begins to climb. Energy is abundant within her and she feels almost young. She drags her heavy self over New England asters, many-flowered asters, blue-eyed grass, Queen Anne's lace, and clover. She is a bulldozer. Plants short and tall bow before her, some springing up in her wake, others recovering more slowly, marking her trail. The red-tailed hawk sees her from his perch high in the upland forest, but he knows she is too big to eat. So does the Turkey Vulture, wheeling high overhead.

She makes almost a hundred meters. Where is that pool? She rests again, surrounded by angiosperms, members of the most recently evolved phylum of plants. If length of the phylum name indicates evolutionary success, then the flowering plants have done quite well: Angiospermatophyta.

Consider the tall buttercup, *Ranunculus acris*, currently askew under the weight of Serpentina's hind leg. In the buttercup, as in all angiosperms, the gametophyte plant has separated into two types, male and female. These reproductive structures have moreover been reduced to complete invisibility, consisting of only a few cells sequestered in the parts of the flower. In the tall buttercup, cradled in its resplendent yellow pentagon of petals, a cluster of female pistils huddle in the middle of the floral head, surrounded by male stamens. At the tip of each stamen is an anther, the home of pollen production. Each pollen grain, barely visible in World Two, houses the male gametophyte. Inside the grain is a three-celled "plant."

At the base of each pistil, a swollen ovary encloses the female gametophyte plant in its center. Beginning as a cell with a single nucleus, the female gametophyte "grows" by dividing its nuclei three times into eight—last vestiges of the old archegonium.

Pollen from the anther of the buttercup might blow across the flower and land on a stigma. Or it might blow across the field to another tall buttercup flower. Or it might travel on the leg of a bee or wasp. But the vast majority of pollen grains end up nowhere, drifting down to the litter layer of field or forest, into a pond or stream, or onto a highway.

But when an extraordinarily lucky grain makes it to the stigma of a tall buttercup flower, the male gametophyte swings into action on the last mission of its short life. One of the three cells of the plant forms a pollen tube that penetrates the surface of the stigma and grows all the way down the style to the ovary, where it enters a tiny pore and proceeds to the female gametophyte, a swollen, elongated cell with eight nuclei. By a cellular mechanism called streaming, the remaining two sperm cells of the male gametophyte now travel down the pollen tube, around the bend, and into the ovarian chamber, where they penetrate the female gametophyte plant. One of the cells fuses with a nucleus there, changing from haploid to diploid. Thus begins the new sporophyte generation—as an embryo. The other sperm cell fuses with two other nuclei to lay the foundations for the embryo's food supply.

The embryo matures inside a seed that develops from a transformed ovary, growing as large a food supply as will fit inside the seed and developing a tough coat to protect it. Again by extraordinary luck, a seed may find suitable ground and germinate the following spring. There will be a new tall buttercup in the floodplain meadow, nodding in the breeze and advertising its sexual needs to passing insects. The sporophyte has said to the gametophyte. "Go about your business of re-production, but let's have no more pretensions to plant-hood."

Serpentina looks around her, smelling water. She advances, tank-like, to a depression in the ground that is barely damp. The pond has all but evaporated in the dry weather. Serpentina looks as though she is pondering the latest advance in the plant kingdom, as symbolized by the blue flag that grows gamely amid some bulrushes and sedges. All three plants are monocotyledons—the latest thing in angiosperms. The seeds of such plants do not have two halves like those of the dicotyle-donous buttercup and other broad-leaved plants. Their seeds have but one cotyledon, and all the plants in this great group are grass-like, including the palms, the lilies, the rushes, and others.

In fact, Serpentina is not pondering the rise of the mono-cots. The past does not concern her. She is as modern as any animal in the Hollow because she is here, because she has sur-vived. She waits a full ten minutes for something to move. Then she moves.

The Labyrinths

Evening, soft and warm, lures the earthworm, *Lumbricus terrestris,* out of his burrow just far enough to expose the first dozen segments. He is handsome to a fault. In World Two his "head" resembles a stack of glistening, brown inner tubes, tapering to a prostomium—a blunt, conical nose for business. Lumbricus points his head upward and waves it in slow circles, nosing the air to see what vegetation might be near. Abruptly he stops and another dozen segments materialize, gliding through the air, then over the ground, to an emerging cottonwood seedling. He slides his prostomium over the shoot, guiding it into the mouth set shark-like on one side of the great nose. He cannot chew the shoot, for he has no teeth. Instead he seizes it with his mouth, prostomium pressing tight, and retracts into his burrow. The shoot slips easily from the ground, and Lumbricus drags it home.

Lumbricus wears his lineage. He belongs to the phylum Annelida, elongate creatures with ring-like, or annular, segments. The annelids began their evolution not millions of years ago but nearly a billion. They abound in the Burgess Shale of British Columbia and in the Ediacara Fauna of Australia. His class is Oligochaeta, "worms with few bristles": Lumbricus has four pairs of bristles per segment. These act as anchor-legs, extending into the walls of his burrow to prevent Lotor from pulling him out. Midway along his body, Lumbricus wears a band-like organ called a clitellum that is deployed during sex, when Lumbricus comes out of his subterranean closet to reveal his true nature as a hermaphrodite. His family name, Lumbriculidae, echoes *Lumbricus* as the type genus. Finally, although he is called *terrestris,* he is quite at home immersed in water. He does not drown because he is equally adept at absorbing oxygen (through his skin) from water as from air. Other members of his family are fully aquatic.

Lumbricus pulls the cottonwood seedling partway into his burrow and then emerges for another foray, shopping for vegetables, fresh or not. The scene about him is busier than it looks at first. He probes the litter zone, a thin layer of last year's leaves, now decayed into darkened skeletal remnants and particles. This layer contains much more: fecal pellets from insects, bird droppings, animal urine, shed skins, arthropod body parts, rotted mushrooms, failed or dormant seeds, feathers, fallen petals, and every detrital element subject to gravity. The litter layer is a vast junkyard rich in recyclables.

Naturally, animals from many phyla abound here. Some chew, suck, or rasp on the special prizes that avid searching uncovers: land snails and slugs (Phylum Mollusca), millipedes, centipedes, mites, and detritivore insects (Phylum Arthropoda), other worms (Phylum Annelida), voles, mice and birds (Phylum Chordata), nematodes (Phylum Nematoda), fungi (Phyla Zygomycota and Dikaryomycota), and even bacteria (Phylum Eubacteria).

Other animals come to hunt: snakes, shrews, and insect-eating birds (Chordata), predatory insects, spiders, and springtails (Arthropoda), and even bacterivore nematodes (Nematoda).

Here also live the common but extraordinary pill bugs such as *Porcellio scaber*. It is not a bug, properly speaking, nor even an insect, but the closest living relative of trilobites like *Phacops rana*. It has a remnant fondness for water, expressed by its love of humid places at the bottom of the litter layer. It feeds on rotted plant remains and, like its distant cousin the trilobite, will curl up when threatened. It will make a pill of itself, coiling its dorsal, arthropod armor around delicate legs and underparts. It belongs to the class Isopoda.

The recycling process begins (and nearly ends) in the litter layer. Complex and simple biochemicals, from cellulose and sugars to proteins and amino acids, are ingested by the browsers who break these compounds down, living on the energy stored in their bonds. In doing this, they must make complex biochemicals of their own before becoming someone else's lunch or dying. Follow the carbon as it flows from plant to animal to animal to animal. Every atom will be respired and expelled as carbon dioxide, or it will be locked for a time in indigestible remnants, joining the soil below the litter.

In the act of dragging a dead leaf to his burrow, Lumbricus senses vibration in the ground nearby, a regular pounding. He abandons his shopping and withdraws abruptly into his burrow to ride the bristle elevator down. Once safe inside, he stops.

He has lived in this vertical apartment all his life, enlarging it as he has grown. Should the food material around the entrance ever cease appearing, he will make a new home by burrowing horizontally some distance and then going vertical again. But tunneling is an exhausting act. The soil that

Lumbricus must swallow to make the journey has barely enough nutrients to pay for the move. There is also wear and tear to consider. His kind of earthworm prefers to stay where it is as long as possible.

To tell the truth, Lumbricus has been feeling a bit dry lately. If his skin dries out to any degree, he will die. He decides, therefore, to ride to the bottom of his burrow, a full meter down, where there is some water. He alternately retracts and extends the bristles of each segment as waves of contraction pass along his body. He slides with difficulty on the drying mucus and compressed feces that line his burrow. The same lining seals him off from the most complicated and extensive habitat in all of Hungry Hollow, a labyrinth of passages and chambers in which water and air flow or rest.

The first five centimeters of his burrow pass through a layer of rich, dark topsoil. Silt, sand, and clay have been suffused with organic material from the litter layer above, darkening this horizon with carbon. The next ten centimeters are a lighter color, with less organic matter. But this layer too consists otherwise of sand, silt, and clay. Such generalizations about what are called the A1 and A2 layers conceal some startling particulars.

The real story of the A-layers is a tale of burrows and passages large and small. They begin in Worlds One and Two, from the simple burrows of gophers and moles to those of crayfish and worms. Not far from Lumbricus' home a colony of *Monomorium minimum,* the little black harvester ant, has excavated an extensive network of vast galleries and tunnels. But what are these compared to the extraordinary complex of tiny pores and interstices made by the random artifice of jumbled grains of sand and silt? In World Five, these endless passages twist and turn every which way, with many *culs-de-sac,* private chambers. They permeate the soil.

This world lies just outside the burrow of Lumbricus, shrouded in the darkness and weight of the soil overhead. No one can see a thing. In the upper horizon, such passages have dried out, invaded by air. A spider mite with glistening amber abdomen might wander down some accidental tunnel, but it would find little, for most wanderers of the meadow labyrinth prefer the watery passages farther down.

Here in World Five, the spider mite is Godzilla to a chamber made of sand-boulders. It cannot enter. Inside, where two boulders meet, some clay crystals have been growing, stacked up in barely visible nano-rafts. A root hair of red clover has invaded the space, and a mass of bacteria, looking like smooth maggots, coat the hair. *Rhizobacter leguminosarum* would invade the root, but the soil has grown too dry and the bacteria must wait. Inside the clover root, other members of this genus are already at work in congenial humidity. They fix nitrogen from the soil, thus making it available to the plant, and receive shelter in return.

Bacteria are the most abundant organisms in any soil. Besides *Rhizobacter,* the labyrinths of Hungry Hollow house *Azotobacter, Bacillus,* and *Nitrobacter,* to name a few genera. To name more than a few species would be difficult, partly because the species concept loses much of its meaning with bacteria. Bacterial species are labile. A bacterium may become a new organism, in effect, by borrowing genetic material from a colleague. It links with the other by means of a pilus, the tiniest of hollow tubes. The strand of genes it receives may endow it with a completely new suite of biochemical talents, changing its life forever.

Fungi are hard at work in the A-layers. A hundred species of *Aspergillus* and *Penicillium* insinuate their hyphae into every space, seeking all that is freshly dead. The hyphae are transparent tubes that work like elongated bacteria. The tip of each tube secretes digestive enzymes that break down cellulose and starch. As new organic matter is assimilated, the fungi reap more energy to grow more hyphae. Like the bacteria, they will let no digestible remnant remain. They are more than scavengers. Some will attack living plants and animals, infesting and infecting. Worse yet, some are predators.

What spelunker would dare to wander these confusing passages? The thread of Ariadne would quickly run out, and no sensible map could be made.

Down, farther down in the particulate soil, groundwater has been drawn up by capillary action. When the passages fill with water, ciliates and rotifers are free to hunt. They motor from chamber to chamber, deadly submarines that graze on bacteria or ingest each other in dark ambushes with no eyewitness.

The most deadly denizen of these liquid spaces is the water-bear, *Echiniscus spiniger,* an eight-legged monster in World Five. It wears a wrinkled cuticle, each leg bears claws, and its snout probes bear-like into every space. Woe betide the rotifer that gets jammed in a corner long enough for the plodding water-bear to reach it. The snout is not a nose but a cylindrical mouth that *Echiniscus* presses against the rotifer. Two sharp stylets dart from inside the mouth, piercing the victim's protective lorica. The rotifer is pinned and begins to leak. Immediately, the pharynx of the water-bear expands, and suction draws the rotifer's body fluids gradually into the water-bear's gut. It belongs to Phylum Tardigrada, the "slow walkers."

One *Echiniscus* has been hunting near the upper limits of the capillary water. It has felt the meniscus so frequently that it assumes a dry spell has come. It is right. It proceeds to secrete a barrel-shaped cask about itself. Made of resistant protein, the wall of the barrel will keep the water-bear wet and alive as it enters suspended animation, to remain within until all is water again—a hundred years if necessary.

The watery passages are also home to legions of nematodes. Some seek drifting plant debris; others hunt bacteria. They thrash and twine from space to space like snakes in seizure.

Lumbricus continues his downward journey, entering a layer of light brown clay. He is master here, for few other organisms have entered. Clay is heavy stuff to mine or excavate. Moreover, clay has a life of its own. At the bottom of his burrow a little water has collected. It rises around his body. Lumbricus takes a drink through his skin.

The pounding vibrations increase until Dianne has passed overhead. She is only occasionally aware of what might lie beneath her feet as she walks the floodplain meadow. But at such times she treads lightly, as though walking on eggs.

Prayer
of the Mantis

In the floodplain meadow, near the edge of the woods, a certain horror lurks in the long weeds. Clinging to a blade of blue-joint grass, *Mantis religiosa* says her prayers, raptorial forelimbs folded in quiet contemplation.

> *Thou hast given me the means*
> *To seize my dinner on the fly*
> *Or crawl. I wait, trusting in Thee.*

She prays that prey will come within seizing range. She is hard to see against the long green grass, for she is also long and green, her covering forewings striated like blades of grass, her legs like stems of grass astray.

Ladybugs do not see her well. At best she is a greenish, indistinct thing ... that is grass, yes, without a doubt. Grasshoppers might see something odd (they know not what) if she leans too far from her blade, so she stays aligned. Robber flies may see her very well, unless distracted by their own prey. She has taken them on the wing. She eats everything.

Her appearance is alien, but not because she is an immigrant. Her ancestors came from southern Europe a hundred years ago in a shipment of fruit. She has spread throughout eastern North America since then, finding a niche not occupied by the more southern species (who share the alien look). Mantid eyes are enormous. Our lady of the grass sees all her prey very well.

Viewed in World Two, the eye of the praying mantis is an expressionless Buckminster Fuller dome, pavilion of the predators. Its gently curving surface is honeycombed with a thousand facets, corneal windows admitting light to analog processors beneath. Each window covers a conical lens that focuses the light from the narrowest of angles onto an arrangement of rhodopsin, a pigment that expels electrons when photons strike it. Window, lens, pigment, and nerve make up the ommatidium, main processing unit of the compound eye. The intensity of light from that narrow angle, embracing but one dab of a pointillist world, prompts the rhodopsin to send a message along the nerve:

> *This much light was seen,*
> *More red than green.*

The signals gather in nerve trunks and speed to the optic lobes below, more than half a mantid's brain. Her being resides here if anywhere. A thousand signals, sorted out and analyzed, may say "prey." Or they may say "keep praying."

Her brain can spot grasshoppers and katydids, though they wear the same disguise as she. The brain knows movement best and tracks the creeping of her grassy colleagues. She will gauge their distance by looking directly at them with her rare gift of binocular vision. Or she will move her head slightly from one side to the other for longer measurements. The brain will decide whether a smash-and-grab operation is in order.

When the bush katydid (*Scudderia furcata*) strays too close, there is a blur of motion in front of the praying mantis. It ends instantly, with the katydid struggling against double rows of spikes in the mantid forelegs, their femur and tibia folded together like a vice. With barely time to mumble her thanksgiving, the mantis turns the prey and cocks her head to inspect it. She will start at the shoulder this time. Her labrum lifts like an upper lip over the soft exoskeleton below the katydid's wing. The mandibles pull wide apart to either side, then cut inward. Their hard, sharp teeth slice through cuticle into the liquid flesh beneath. Once, twice, and again. Pieces are coming off. The labial palps behind the mandible pass each fleshy fragment over what passes for a tongue, and the mantis gulps it in, down the long thorax and into the crop.

When she has eaten nearly through the katydid, the mantis takes stock with her thousand ommatidia once again. If she keeps eating, the body will fall into two parts and become unmanageable. She knows this. Perhaps she learned it.

The katydid's head is joined massively to its thorax, so the mantis ceases to eat her way through, changing her angle to eat toward the head. She eats the eyes, and even as she eats the brain, the legs of the katydid twitch, thrash for a second, then twitch again. Insect limbs are run by other, smaller brains further back. What news do they have of the disaster? Do they know that the head—with eyes, brain, labrum, mandibles, and palps—is already gone? Will they know when their turn comes?

The mantis cannot eat now. Her crop is distended and she rests on the blade of grass. Within her long, capacious abdomen, fifty unfertilized eggs wait in ducts for their mother to choose the place and time to deposit them. They have developed for two weeks now, but they have only half their genes. The other half will come from sperm that is already near at hand. When they are born, they will pass a small pouch called the spermatheca. Here, their mother will have a supply of sperm, offerings from their father-to-be, a mantis who-once-was.

How can *Mantis religiosa* expect salvation when she has eaten the head and thorax of her husband? She may not re-member, but it happened like this. When her tiny eggs were ready, she gave off a scent, a chemical perfume that spread on the summer breeze and wafted by a male near Hungry Creek. Having got wind of her pheromones, he flew clumsily, in the manner of a grasshopper, to her patch.

He found her blade of grass and climbed the stem cau-tiously. She felt his vibration and turned herself around to look down the stem. When she saw him and he saw her, it was not love at first sight, but a kind of insect terror. His eyes were as good as hers, and he could see that she was looking at him, two false pupils aimed in his direction. He froze, then retreated instinctively down the stem until he could no longer see her clearly. The decision to stop was hard-wired. She would not see him clearly, either.

The courtship amounted to a slow dance, the male creep-ing and creeping up. The female had turned her back to cling once again, surveying her upright world and continuing her devotions. Perhaps she had the male in mind; perhaps she forgot. For his part, the male moved more slowly as he approached her from behind, praying she would not notice. His spermatophore, a tubular purse inside his abdomen, was full. He dared not tremble now but stalked his love as he would an overgrown grasshopper that might take flight at the slightest sign.

He leapt.

There was confusion, a swinging of two bodies, one seizing the other, as the female held on by a single leg and the blade of grass bent low. The male held her by middle and hind legs, crouching flat against her back. Her head swiveled right and left to see what had clasped her so rudely. The mantis is not used to being prey, especially to one smaller.

The male found her genital pore and turned the opening of his sperm duct inside out to make a temporary penis. He allowed no motion from her of which he was not a part as he deposited the spermatophore, a ropy bag of sperm, into the pouch beside her genital pore. The process was slow. The female ceased her struggle, biding her time.

When at last the deed was done, the male relaxed his hold. Gone was the urgency provoked by the magic smell. He would not linger with this dangerous beast. He spread his leathery forewings to be gone.

Alas, she was too fast—or he too slow. She seized his head as he dismounted, pulled it unceremoniously to her mouth, and ate it. Then she ate his thorax. That was enough. She released his body. It clung momentarily to the grass stem, then slipped off and fell to the earth below, where scavenging ants would soon discover it.

Ever since then, his sperm has awaited the fifty eggs. The blessed event will come late tomorrow morning. The eggs will pass the spermatheca to be fertilized one by one. They will continue into the genital pouch where a special froth whipped up by female glands will surround them. The female will lay them, one by one, stacked in a long raft, surrounded by foam.

The foam will harden, and the eggs will overwinter in suspended animation. In the spring they will develop in their secure house, maturing into tiny replicas of their mother or father. Emerging, they will molt immediately to make their spindly way along the smaller grass and mosses to hunt springtails, mites, and ants. Here they will begin tiny prayers that, if answered, will bring all fifty children through five successive molts to adulthood, complete with wings and sexual organs.

But most will not make it, nor will they understand the wisdom of their destruction, prayers notwithstanding.

THE HYDRAULIC PLANT

There has been no rain for many weeks. On higher ground, where the floodplain meadow meets the forest edge, the orchard grass (*Dactylis glomerata*) is under stress from drying. A young toad pauses by the stem on its way to lower ground where, perchance, it will find water. A faint sigh from the stem, followed by a popping noise, startles the toad. Fearing an unknown predator, it hops quickly away. It knows nothing of plumbing.

In World Three, where stems soar a kilometer into the sky and roots dive hundreds of meters into the parched soil, details of the crisis emerge. What little water the plant has is packed away in cells for turgor, the pressure that holds the factory up.

In World Three, the stem of orchard grass is a cylindrical green tower as thick as a silo. Far above, the leaves amount to five immense solar collectors that arch upward from the tower. Inside the collectors, the looms of light have nearly ceased to work. They have much light to weave by, but no carbon dioxide for thread.

Within each collector, a spongy layer of chlorophyll-laden cells occupies the middle space, permeated by stale and humid air. Outside the collector, waterproof epidermal sheeting is studded with stomata, portals of exchange between the air within and the air without. The stomata have all closed, for if the humid air within the collector escapes, the cells must bleed a little of the precious fluid into vapor. The collectors have also curled into tubes, enclosing the stomata as a further protection against drying. And just as water vapor cannot escape the collectors, carbon dioxide cannot enter. The factory is shut down.

If the plant cannot produce new energy for itself, it can only respire, using what starch is on deposit. It runs on a budget tighter than that of any business. It will make withdrawals from its own starch accounts, converting them to glucose and then to ATP dollars.

The plant cannot borrow from the Central Bank of Hungry Hollow. If it depletes the starch before relief arrives, it will go into catastrophic, irrecoverable failure. Tissues will collapse, wilting. The plant will brown into death.

At its base, the tower plunges underground to its foundation of roots, a basement of cellular plumbing that extends outward and downward, ever branching.

The science of groundwater recognizes three zones in every soil that is neither desert nor swamp. Water saturates the lowest zone, whether solid clay or a matrix of loam. From this zone, water creeps up by capillary action into the particulate soil. A molecular web works within every space to draw the water after it. The capillary zone reaches upward a few centimeters, and then gives way to a dry, or xeric, zone that is slightly humid, perhaps, but devoid of usable water. All three zones have retreated lower with the recent dry spell. The capillary zone has

"The plant cannot borrow from the Central Bank of Hungry Hollow."

sunk below the deepest roots of the plant. Each root hair fights a losing battle to draw in microscopic bits of water from the interstices between grains of soil. Soon the plant must make a decision, in effect. Should it thrust its deepest roots still deeper? To do so might empty the starch accounts.

The rule of groundwater is this: Deep within the soil it will creep, a slow, invisible sheet, along gradients in the land, settling always for the lowest venue. Thus the groundwater of Hungry Hollow has deserted the upper flood plain for the lower, seeping down the gentle slope toward Hungry Creek. Near the creek it levels off below the soil surface, but no lower than the creek itself. Here, an abandoned water-course hints at its former route, stones of the old stream bed haunting the weeds.

When the young toad nears Hungry Creek, it enters the old runnel and pauses once again, quite by accident, near another plant of orchard grass. Here it can smell water everywhere. Beneath the abandoned bed, the roots of the grass mingle with the saturation zone.

This factory resembles the one farther up the slope but has a freshly painted look. The plant is in full production because its stomata are open for business. Carbon dioxide readily permeates the collectors, and humid air from within leaks out. Fresh water streams up from the roots and enters every producing cell to be cleaved by the power of light in the chloroplasts. Oxygen leaks out of the stomata to be breathed, sooner or later, by animals and other organisms. The hydrogen is bonded to the carbon of CO_2 in World Eight, and the resulting carbohydrates are conducted away, for storage, by phloem pipes.

As molecules of water leave the spongy layer within each collector, the tiny crevices between cells, from which the water creeps, grow deeper. Acute angles in the crevices make for large capillary forces. The tension of the web that spans every water–air interface pulls on the smallest supply pipes of the venous system nearby, like a child sucking on a straw. These small pipes must draw on the larger ones to which they connect. The tensions that begin in the spongy layer of the collector multiply a thousandfold as they merge toward the stem.

The largest pipes from the solar collectors join the xylem system inside the tower. In monocots, such as grass, the xylem is arranged in bundles—great vertical mains of clustered conduits that run the height and depth of the tower. They are not primitive pipes, like those in human factories, but the highest of high-tech. Once they were separate, cylindrical cells, but their have thickened walls with lignin and cellulose, and they have fused end to end. Perforations in the end walls have joined the contents of successive chambers. Pits in the side walls, as well, join adjacent cells by one-way valves. The xylem cells that once inhabited these chambers have long since died, their life sacrificed to the transport of water.

To pull upon water in a confined space, as in a straw, is to pull upon a thread that will not break unless the pull is enormous. In the downslope orchard grass, water flows uninterrupted from root to leaves. It makes its way by a hundred channels up the vascular bundles, always drawn by evaporation in the leaves, until it enters them, running along veins to the spongy tissues. By nature's calculation, only a fraction of this water is needed for photosynthesis. The rest transpires into the air.

In the upslope plant, things are not going so smoothly. Unabated, the multiplication of forces pulls each thread of water toward the top, yet the roots cannot supply more from the bottom. The tension mounts until one of the threads can no longer hold. A micro-bubble of air in one of the xylem chambers is drawn out by the tension, vibrating as the remaining water percolates out of the chamber through the tiny holes in the end-plate. The stem sighs. Elsewhere a micro-bubble expands more quickly, exploding into a xylem chamber. Both times, the mini-holes in the sophisticated pipe-chambers prevent the bubbles from spreading. The valves in the sidewall pits snap shut. Isolated, the break cannot spread.

The orchard grass in the old runnel neither sighs nor pops. The pull on the threads of water runs down the stem to the roots, where the bundles divide, then divide again, roots branching into rootlets. In the Stygian dark of World Four, each rootlet shoulders its way into the soil, root hairs invading the watery labyrinths. In the interstices they absorb water passively or they actively pump it in.

Thus the orchard grass in the old stream bed transpires water from the ground to the air, from root hairs to leaves, by a system of tubes, pipes, valves, and conduits as complicated as the circulatory system of any animal. Other plants by the creek also transpire. The trees in the forest transpire, for their roots are deep.

But the other Orchard Grass has only a day to live without moisture. Its fate has already been decided in the Gulf of Mexico.

The Storm

Humid air has invaded Hungry Hollow from the south. The atmosphere is still and the cicada sings. Its buzzing rattle fills the breathless air between the trees and permeates the hot haze that cloaks the valley. The song echoes like burnished brass, like a gong struck by the sun.

Choirs of solar photons descend, singing the visible frequencies. Some photons collide with molecules of air. Low notes are absorbed, and the molecules spin and vibrate to the same tune. High notes ricochet, striking plants or soil, re-emerging an octave lower, in the infrared. Molecules of water vapor absorb them, sympathetic to their vibrations.

The haze comes from the creek, from the ground, and from plants. It has been stripped in micro-sheets from the slow, tepid water of Hungry Creek. It has been wrung from the last reservoirs below the dry ground. It has flowed up the tiny pipes of innumerable roots to stalks and trunks, then up to leaves, transpiring. In the greenhouse of the valley, the air grows warmer without limit.

Where is relief from this cloying humidity heavy with the smell of plants? The answer is inscribed by the air itself against the backdrop of distant trees. They shimmer as the vapor rises in front of them. Before any water may fall from clouds, it must first ascend to make them.

The water rises from Hungry Hollow in airy parcels of hot vapor that coalesce among the plants in small, ephemeral pockets. These ascend slowly, a gaseous paisley of miniature hot-air balloons, merging as they rise. They float heavenward meters above the valley floor, then tens of meters.

As they rise, the air pressure slowly falls and each balloon expands. Its molecules draw farther apart, slowing the dance of heat. The temperature falls in the balloon, but the heat remains, dispersed within a larger volume. As the balloon cools, the water vapor creeps ever closer to the dew point, where it will condense.

The balloons rise higher and the air around them grows cool. If the air is stable, it cools less quickly than the balloons as they rise. At a certain height, the balloons reach the same temperature as the surrounding air and lose their buoyancy, coming to a gentle rest and dissipating. But when the air is unstable, as now, the air cools more quickly than the balloons as they rise. They bobble steadily upward, merging into thermals that lift the turkey vulture as it searches for death in the floodplain meadow below.

At the critical altitude hundreds of meters above Hungry Hollow, thermal packets reach the dew point and become visible. At this temperature and pressure, the water must condense

"Before any water may fall . . . it must first ascend . . ."

to form tiny droplets. They would form miniature clouds of their own if there were not already a cloud there. They are lost just when they would be seen.

Their heat is not lost. When the water condenses, more heat is added to the air surrounding each droplet, slowing the condensation. All the warm, moist air within the cloud is still lighter than the air without. Thermals merge into a vast, irregular column that rises inside the cloud. This column cools as it expands, condensing as it goes. It ascends to the top of the cloud, emerging in a distant cauliflower against a deep blue sky. It is fair-weather cumulus, drifting slowly away from Hungry Hollow to the east. The sun comes out.

The red-tailed hawk spies a short-tailed shrew creeping near the rotting log by Hungry Creek. It flaps away from its perch high on a dead black cherry at the forest edge. It rides the miniature thermals out over the meadow, then dives from out of the sun, wings quivering with a hundred minute adjustments, pressure building on feather airfoils. At the last moment it levels and seizes the shrew, which squeaks and then promptly dies of fright.

There is a new cloud to the west approaching the Hollow. It has grown from fair-weather cumulus into something else. It has become a convective organism that swallows the warm, saturated air from below, sucking it into a vertical gust. Within the rising columns, droplets merge into small spheres of rain that grow heavy and rise less quickly. The cloud top has risen almost to the top of the troposphere—to a layer of cool, dry air that will arrest all convection. The cloud top flattens against this layer, forming an anvil. It is towering cumulus.

From its new perch on a fence post on the edge of the corn field, the red-tail looks up from the shrew under its claw at the sound of distant thunder. In the approaching cloud the rain-drops have grown heavy, descending through the convective core. There will be a storm.

The light grows peculiar as the cloud approaches. There is a drop in air pressure. The cloud creature has mutated into a beast. The convection zone enlarges into a crescent mouth that vacuums up new air along the advancing front. Behind the mouth rain falls, dragging cold dense air from the cloud, air that spreads over the obdurate ground and rushes out of the front, a large, invisible tongue. A sudden cold gust nearly blows the red-tail from its post. With the dead shrew in one talon, it beats toward the wood and sounds a falling scree-note to its mate.

The cold dense air slides under the warm moist air of the meadow and woods. Impatient of slow balloons, it sucks all upward into its invisible mouth. Within the storm front, the heat of Hungry Hollow rises, feeding the beast. Behind the mouth it falls, cool and rain-laden. Between the two, air is in turmoil, sometimes rising, sometimes falling. Hail forms here, rain frozen by indecision.

Crack! Boom!

The form is intense, blue-white. An erratic, meandering river of electrons joins cloud to ground for less than a moment. Under the pressure of 500,000 volts, enough electricity has just entered the soybean field north of Hungry Hollow to light a medium-sized city for a minute. The electrons dissipate into the soil, but not without harm. More than a dozen earthworms are electrocuted, along with several thousand nematodes.

The river of electrons is only a few centimeters wide, but it heats the air to 35,000 degrees Celsius. Baked and ionized, the air is shocked into an expansion that does not stop when

114

the lightning ends. It spreads outward as sound. Waves from the lower bolt arrive as a sharp crack that startles the red-tail and its mate, perched on the underbranches of a white oak.

The crack wakes Lotor, who stirs to open his eyes and raise his head to the doorway of his hackberry apartment. He sees a figure running up the path across the valley, ascending to the suburb. She holds something over her head.

From higher sections of the bolt, up in the cloud, the sound arrives later, smeared into a boom. Lotor knows about storms. He scratches fiercely at a flea on his neck.

The history of the lightning bolt must be written in milliseconds. It began with electrons in the lower cloud, transported there by hail and rain. They formed an immense volume of collective charge with nowhere to go, insulated from the ground by air. A dearth of charge in the ground invited a gradual breakdown of the insulator. Attracted to the ground, a tentative zig of electrons flowed several meters down a tiny channel of ionized air below the cloud. Microseconds later, a zag extended the channel many meters more. The attraction grew with proximity. When the zig-zag step leader met the ground, electrons in the last step rushed into the ground, then those from the previous step, then those from higher up. The return stroke of successive downward motions opened the channel. Electrons from the cloud crackled down in their myriads.

More lightning followed, much of it within the cloud. The convection had separated electric charges into regions that would repeatedly re-connect with white-hot ropes as their burden or scarcity of electrons grew to unbearable levels.

The rain came to Hungry Hollow like a blessing. Water that had been carried aloft in parcels far to the west now descended in drops on the valley. It fell in the forest, soaking through last year's leaves and cracked earth. It fell on the parched meadow to replenish the groundwater and feed the roots of orchard grass. And when the ground near Hungry Creek was soaked, drops spread over the surface, flowing by fits and starts into the current.

As the creek rose, it slowly turned brown with run-off from the field to the north. The air grew cool and steamy, and Lotor awoke again. He thought he might go a-worming.

Abundance

It is two days since the storm. Dianne has been wandering the floodplain meadow, marveling at the renewal of life. She stands now in the midst of goldenrods, lost in their nodding dance, mesmerized by the complexity of it all. Stiff, straight stalks bear a ladder of lance-shaped leaves ascending to a spray of yellow fronds, each a string of tiny daisies that waves before her eyes, level with them.

She feels delightfully awash at such moments, astonished by new habitats at every turn. She cannot fail to see the ants that crawl the stem, the lady-beetles that prowl the leaves, the ambush bugs that hide in the flowers. She cannot miss the bees, wasps, hornets, and butterflies that come to call for pollen and nectar. The tall goldenrod is not just a plant but a vertical world with dramas playing every moment. The habitats nest, each within the next, like a fractal. The complexity drags her into a vortex. It demands too much. She cannot keep up.

The scientist within must grasp the place as a whole. It is a compulsion. She thinks that by learning every plant in the meadow, she will achieve the yearned-for perspective. But her training has not equipped her for the project. As an undergraduate, she learned only general principles supported by a handful of well-chosen species. As a graduate student, she plunged into nutritional studies of the ciliate *Tetrahymena pyriformis*. This does not help her identify plants, either.

She has never, before this summer, directly confronted the astonishing variety of life. She is, she admits, an amateur, the wildflower book in her hand her only guide. So far she has identified some fifty plants in the meadow, but now the project is blowing up in her face. There are too many species, and the process of identification becomes harder, not easier, as she comes to the difficult groups like the grasses and sedges, not to mention the goldenrods.

Dianne has decided that the goldenrods around her are tall goldenrod, *Solidago altissima*. She is right this time. What other species of goldenrod are about? She looks around her, seeing only tall goldenrod. It is very abundant in the floodplain meadow. She sighs. Perhaps she will try another genus.

There is a new plant at her feet. It comes only to her knees, but it bears several showy, purple flowers that speed identification. Her wildflower guide says New England aster. *Aster novae-angliae* looks quite different from goldenrods, but it belongs to the same family of composite-flowered plants, the Compositae. The aster is presently receiving just as many visitors as the goldenrods. It is the first one she has seen while wandering the meadow, and she wonders where the others are. She is on the threshold of a great discovery.

Dianne will not discover that there are exactly three New England asters in Hungry Hollow at the moment. Her discovery will involve a more general understanding, a pattern that will shape her wildly burgeoning knowledge.

She walks the length of the meadow twice, looking for the New England aster, but finds no more. She wonders if it's a rare species and consults her guidebook: "common throughout eastern North America." It certainly is not common in Hungry Hollow, at least not at the moment. The great discovery edges closer when she spies a crown vetch, the same one she catalogued before the great storm. Where are the other crown vetch, she wonders. She goes looking and, after much effort, locates another. Obviously not abundant. She goes on to find just two enchanter's nightshade, one heal-all, three daisy fleabanes, and two English plantain. Enough!

However abundant these species are elsewhere, they all have definite abundances in the Hollow. Some are very abundant, yet some are very rare, in this local quarter. The great discovery is at hand. "Actually," she murmurs to herself, "there are only a few abundant species, but a lot of rare ones."

Out of nowhere, some lines of verse comes to her. Is it the Hollow?

Species here and species there,
Few are common, many rare.

What about insects? She can only think approximately; she doesn't know her insects yet. But since nine this morning (it is now noon), she has seen a lot of wasps, a lot of ants, a lot of sulfur butterflies, a fair number of honeybees, and several bumblebees, but only a few ambush bugs, one hangingfly, one ichneumon wasp, one housefly (or whatever it was), one small copper butterfly, one mantis, one katydid, three grasshoppers, two dragonflies . . . Hmmm. Maybe.

The most manageable numbers will come not from small organisms but from the largest. Consider the trees, oh Dianne!

That afternoon and for the next few days, she switches from herbaceous plants to woody ones. Although she gets a bit mixed up between red and white ash, and between blue beech and juneberry, she systematically surveys the forest above the meadow, counting each species. The better she gets, the faster she goes. For good measure, she throws in the trees that grow out on the meadow and along the creek.

That night, working late in her study, she lists all the trees and shrubs she has counted, along with the number, or abundance, of each:

American beech	*Fagus grandifolia*	24
Ironwood	*Ostrya virginiana*	20
White ash	*Fraxinus americana*	19
Sugar maple	*Acer saccharum*	15
Red oak	*Quercus rubra*	13
Hawthorn (all species)	*Crategus* spp.	11
Shagbark hickory	*Carya ovata*	10
Buckthorn	*Rhamnus cathartica*	9
White oak	*Quercus alba*	8
Blue beech	*Carpinus caroliniana*	8
American plum	*Prunus americana*	7
Witch hazel	*Hamamelis virginiana*	6
Juneberry	*Amelanchier laevis*	5
Red ash	*Fraxinus pennsylvanica*	5

Silver maple	*Acer saccharinum*	4
Highbush cranberry	*Viburnum lentago*	4
Flowering dogwood	*Cornus florida*	3
Pawpaw	*Asimina triloba*	3
Hackberry	*Celtis occidentalis*	3
Basswood	*Tilia americana*	2
Largetooth aspen	*Populus grandidentata*	2
White elm	*Ulmus americana*	2
Black willow	*Salix nigra*	2
Black ash	*Fraxinus nigra*	2
Grey dogwood	*Cornus racemosa*	1
Black cherry	*Prunus serotina*	1
Black gum	*Nyssa sylvatica*	1
Chinquapin oak	*Quercus muehlenbergii*	1
Swamp white oak	*Quercus bicolor*	1
Eastern cottonwood	*Populus deltoides*	1
Peachleaf willow	*Salix amygdaloides*	1
Sycamore	*Platanus occidentalis*	1
Pignut hickory	*Carya glabra*	1

She becomes excited. The trees seem to tell the same story as the plants, but now she has the numbers. Only the big picture is lacking. She makes a histogram, a chart that inverts the data. Next to a column of numbers from 1 to 24 (the greatest abundance), she draws a row of squares beside each number. Each square represents one species that has that abundance.

Abundance	Number of Species
1	❑❑❑❑❑❑❑❑
2	❑❑❑❑❑
3	❑❑❑

4	❏❏
5	❏❏
6	❏
7	❏
8	❏❏
9	❏
10	❏
11	❏
12	
13	❏
14	
15	❏
16	
17	
18	
19	❏
20	❏
21	
22	
23	
24	❏

The length of the first bar shows that there are exactly 9 species with just one tree in Hungry Hollow. There are 5 species with two trees each and 3 species with three trees, and. . . . The lengths betray a downhill trend:

$$9 \quad 5 \quad 3 \quad 1 \quad 2 \quad 1 \quad 0 \ldots$$

The seventh number is a zero because, as it happens, no species of tree has exactly seven members in Hungry Hollow.

Beyond this, zeros and ones alternate unsteadily, the ones becoming gradually sparser.

When she turns the chart on its side so that the numbers run along the bottom of the figure, Dianne sees the shape of the distribution in standard position. The numbers of species in each category fluctuate, to be sure, but the overall progression seems remarkably clear. There is a curve that slopes sharply down and then evens off, getting closer and closer to zero, on average.

Dianne has just discovered the J-curve. Shaped like a backward letter J, it falls rapidly at first and then levels off. Field biologists know the J-curve well. Sometimes it is steep, as now; sometimes it is shallow.

There are many species of trees and plants in Hungry Hollow that belong to the phylum Angiospermatophyta, but there are far more animals in Hungry Hollow that belong to the phylum Arthropoda. In fact, there are far more insects within this phylum. Here too, the rule that Dianne whimsically formulated applies, at least in relative terms.

Few are common, many rare.

It applies not only to the insects that live among the goldenrods and asters but also to those that live under bark, hunt in the dark, creep on the ground, or whirl busily on the creek.

The most common species of insect in Hungry Hollow at the moment is the Allegheny mound ant (three mounds with a total of 1,284,129 subjects and royals). The next most abundant is a chironomid midge (seven swarms with a total of 867,035 aerial dancers). Then comes the tent caterpillar (58 tents with a total of 355,828 diners). As we move from species to species in the direction of lower abundances, the numbers decrease more slowly. At the "rare" end, Hungry Hollow has 17 seventeen-year cicadas (a coincidence), 10 white-tailed skimmers, 7 spiny tree hoppers, 4 walkingsticks, and just one water scorpion. In other natural places that have J-curves of their own, some of the latter species are more abundant, relatively speaking.

In Hungry Hollow (and in the other places), species change in abundance every year. All things considered, increase or decrease in number is pretty much a chance affair. Conditions too numerous to know, and interactions too subtle or complex to grasp make fluctuations not only inevitable, but also largely unpredictable as to direction. This year's rare species may be abundant next year, or it may fail to show up altogether.

If it loses a species altogether, Hungry Hollow will have to wait for it to be resupplied from a neighboring habitat. Unfortunately, there are fewer such neighbors every decade. The wait may be a long one. And what if the species exists only in Hungry Hollow? Enter the spiny tree hopper.

The majority of insect species in the world are unknown to science. Although most of them live in the tropics, many unknown species await discovery in North America. An entomologist who stumbled upon this elegant creature might write the following informal description:

> **Proposed name:** "Spiny Tree Hopper" *Thelia solis*
> **Description:** A large (1.2-cm) hopper (family Membracidae) with the shape of a buffalo. Head subterminal, pronotum prolonged anteriorly in two sharp spines, wings with two bright blue bands in the form

of an X, reminiscent of Confederate flag, other features characteristic of genus *Thelia*. One specimen only, Hungry Hollow forest.

The entomologist would then add the date, vowing to return the following year to find more specimens. Alas! It will be in vain.

Dianne, who has only begun to explore the forest systematically, does not know about the spiny tree hopper. But she finds the J-curve very instructive, not to say alarming. It tells her not to worry less about low abundance but to worry more. If it is natural for many species to have low local abundances, she will hope that they have other homes. Next year, however, planet Earth won't be quite the same, with humans none the wiser. The spiny tree hopper will be gone forever.

In the Forest

In the forest, darkness is fashioned into great columns and arches. In the forest, silence becomes a distant thrush, the fall of an acorn. Sunlight glorifies the crown but descends only in small shafts, brightening the life of a sapling, falling on fallen branches, or briefly blessing a wood frog. There is the musk of old lignin and humus, the scent of dead leaves and fungi, the faint perfume of small flowers. Clocks run slow. A year in the meadow takes a century in the forest. And rulers are two short. The plants are ten times as tall.

We are too small and we live too short a time to see the forest boil with trees, like the top of a cloud. We cannot watch the spectacle of old kings crashing continually to earth, dead trunks and branches sinking into the earth, leaves melting, brackets plodding over dead bark, mushrooms hopping like mad elves. We miss, too, the way the royal brood seize the sun in their arms, stretching toward the narrowing sky. We do not see the forest work.

"Sunlight glorifies the crown but descends only in small shafts . . . , briefly blessing a small frog."

From a distant road, the top of the old Hungry Hollow forest looks flat, a horizon of crowns. Closer, it becomes the uneven struggle among kings and princes of the upper realm: beech, white ash, red ash, white oak, red oak, black oak, basswood, sugar maple, and black maple. Courtiers of the understory catch the royal leavings: ironwood, juneberry, blue beech, and witch hazel, down to the humble spicebush and maple-leaved viburnum.

The trees that descend the slope toward the floodplain meadow cannot aspire to such heights, disadvantaged not by breed but by footing: shagbark hickory, yellow birch, red and silver maple, black ash, elm and largetooth aspen. Here the understory turns to buckthorn and highbush cranberry, hawthorn, and flowering dogwood. At the forest edge, the open glades are choked by shrubs and struggling saplings. Vines of river grape and Virginia creeper grapple for sturdier wood. Meadow plants invade the ground. There is more light.

On the floor of the forest, last year's leaves have turned dark and curly, disintegrating. In the litter, grey squirrels, fox squirrels, and southern flying squirrels hunt for nuts and seeds. Least shrews, short-tailed shrews, and starnosed moles seek insects. Garter snakes, ringnecks, and ribbon snakes hunt for insects, amphibians, and each other. American toads, grey tree frogs, wood frogs, and spring peepers hunt for insects.

In the litter, weevils, bugs, beetles, moths, gnats, flies, ants, wasps, hornets, and bees hunt for decaying plant matter, fungi, blood, dead flesh, nectar, seeds, nuts, or each other. Hunting spiders wander here, with daddy longlegs, centipedes, millipedes, and mites. There is hardly a square centimeter of litter that does not have its arthropod or ten.

The forest floor is a miniature plain of dead leaves graced here and there by shade-loving forest plants that succeed each other with the season: skunk cabbage, trout lilies, trilliums, and wood anemones in the spring, then jack-in-the-pulpit, Solomon's seal, false Solomon's seal, running strawberry, yellow mandarin, wood aster, tall white lettuce, and many other herbaceous plants, including the ghost-white, parasitic beech drops and Indian pipe, which live on nutrients stolen from the roots of trees by fungi. Ferns unfold in moist hollows within the forest: Christmas fern, rattlesnake fern, ostrich fern, sensitive fern, and bracken. Mosses encrust old logs and the bases of trees.

There is something strange about the soil beneath the litter, the insects, and the plants. Most forests have a layer of composting humus directly below the leaf litter, but our forest does not. The soil is bare clay, grey and dark, smooth and featureless. It has been this way for nearly ten thousand years, since a vast but shallow glacier ground its way from the north to Hungry Hollow and a little beyond and then retreated with the great warming. But it paused over the hollow, leaving a shallow moraine of silt and clay, a broad upland that supports fields to the west and Whispering Pines to the east. Hungry Creek has carved a valley through the old moraine, scoured by meanders to its present broad expanse.

Where does the litter go? When leaves and twigs have decayed to particle size, they do not join the humus, the A1 layer. The forest has no humus now, it never did, and it has never been the worse for it. Where does the litter go? It cannot sink into the obdurate clay, for only water molecules can penetrate the horizontal micro-crystalline flesh. The water must therefore carry the particles away, perhaps across the floor of the forest and down the slope to the meadow.

But it doesn't.

There was a clue to the mystery—a clue that was distributed over the entire forest floor before the rain storm that swept through Hungry Hollow a week ago. In the preceding drought, the clay had dried out, breaking into a network of cracks. The cracks widened, some of them yawning as wide as a centimeter. When the rain fell, the water immediately began to flow into the cracks, carrying all the recent particulate matter with it. By the time the cracks had filled with the humic soup, the floor was clean once again below the litter. Then the clay began to absorb the water and, over the space of days, swelled again until the cracks, big and small, had all closed. A thousand little mouths had slowly opened and then shut, eating the would-be humus and digesting it directly into the A2 layer.

The clay is very dark with dissolved organic matter. The darkness extends to the depth of the widest cracks, some fifteen centimeters down, where the original clay, light brown in color, takes over. The brown clay is several meters deep, giving way at last to a meter or two of silt and sand that lies directly atop the old Hungry Hollow formation. It is a typical smallish moraine, the mechanically sorted burden of glacial melting into a temporary lake.

Because the dark, nutritive layer is relatively shallow, the lateral roots of trees remain close to the surface, breaking through the litter here and there, a gnarled wrestling of woody legs. Fungi abound, their mycelia probing the clay for remnant cellulose and lignin. Mushrooms and boletes, polypores, stars, buttons, crusts, cups, corals, balls, tongues, phalluses, and fingers poke through the leaves, the unseemly fruit of millions of kilometers of invisible hyphae.

Throughout the upland forest and its meadow arm, other fruit are falling. Seeds, nuts, and berries pelt the floor at random times, squirrels alert to the sound. Less than 1 percent of the fruit ever germinates. The rest is food for animals, arthropods, fungi, and bacteria. Seedlings have a long way to go, and most die in their first year of growth. Of the survivors, only a fraction make it to breast height.

In the struggle for light, every young tree runs on a tight budget. Leaves cost ATP dollars, but sunlight brings more income. Will revenues exceed expenditures? If so, the tree can grow, expanding its base of operations below the ground, investing in new branches and leaves. On sunny days, patches of light migrate eastward across the forest floor as the sun tracks to the west. A patch of light that falls directly on young leaves, even for 20 minutes, may produce as much revenue as an entire day of indirect light.

There is a large element of luck in all this. Seeds that fall on choice soil may nevertheless find themselves shaded out after germination, that precise spot being a no-grow zone. It may also happen that a young sapling does very well in the early stages because it happens to lie on the track of a small sunny patch that visits it every day the sun shines. But as the sapling becomes larger, it outgrows the patch, and no new spotlights wander across its new leaves. The sapling lives on for a few years, barely growing and gradually weakening. In the end it dies and becomes a spindly vertical corpse, a feeding ground for fungi and beetles.

In the general story, the rate at which young trees finally reach a spot in the canopy or the understory exactly equals the rate at which old trees die. Light rules. Any time that more light becomes available, more trees rush to fill the gap through which the light arrives. There is dynamic stability.

The death of a tree is seldom sudden. One by one the branches die, leaving twisted forks in the sky. The remaining branches try to make up the deficit, even as the trunk becomes a vertical extension of the forest floor habitat. Fungi and insects mine its bark and bore into the xylem. They nibble at the cambium until, at some point, the tree is girdled. Or the tree becomes weakened by the loss of heartwood and the burden of its wet rot, crashing without notice to the ground. It may live on in that state, new twigs growing upward once again to the light, fed by the few roots that still remain in soil. Dead, it crumbles over the years into a temporary soil, doing its final service as a nurse-log in which new seedlings may root.

A tree that is already dead may stand for years, however. In either case, the dead or dying tree becomes a valuable resource, a habitat in its own right. Lotor's hackberry is not there yet, but it will be soon.

The forest of Hungry Hollow is not so dark as it used to be. More light enters these days, encouraging the herbaceous plants and prompting the growth of invading, shade-tolerant woody plants such as hawthorns, high-bush cranberries, and buckthorns. Some trees are dying before their time. A red oak, recently crowned, has already begun to lose branches. A seemingly healthy beech fell suddenly last week with a groaning crash that no human heard. All the black cherries but one are already dead. The sugar maples look sparse upstairs, as though they have lost their zest for life. There are only three flowering dogwoods left, the ones growing near the forest edge.

In every case, a particular pest or pathogen is to blame. The beech rotted out rather quickly after being attacked by a scale insect called *Cryptococcus fagi.* The insect brought with it a fungus, *Nectria coccinea,* that invaded the weakened tissue, ultimately killing the tree. The dogwoods were killed by a fungus aptly called *Discula destructiva.* The loss of trees within recent decades is not coincidental (as though the insects and fungi had suddenly decided to gang up on the trees). There is a weakness pervading the forest. It has visited the trees *en masse,* making it harder for them to absorb nutrients, altering chemical balances within, poisoning their metabolism to the point of vulnerability.

The forest is still a beautiful place, but it is not what it was. In ten years, it will be recognizably worse. The forest is slowly dying. New chemicals are making their presence felt: falling with the rain, drifting on the wind, or floating in the creek. There are acids, heavy metals, poisonous radicals and ions, and even nutrients that, in superabundance, act like poisons. Their combined effect is subtle, as though conspiring to appear innocuous. Humans who deny the effect will always be able to make a case, even after the forest has vanished.

The Art
of Decay

A rounded, leathery shelf clings to the base of a dying beech. The upper surface has been painted in concentric half-rings of ochre, burnt sienna, orange, and taupe. The lower surface is creamy white, punctured everywhere by minute pores. *Ganoderma applanatum* has been sculpted in a tough medium that can barely be broken with a hammer. It is the artist's conk, a polypore. Scrape a pointed twig across the lower surface and the track emerges like magic, a dark line of bruised flesh. Draw the diagram of a hypha, greatly enlarged. The elongated cylinder, walled off into individual cells, will appear within a minute.

Over the course of two years, the fungus artist's conk has woven itself from its own hyphae. It has added increments to its circumference, invisibly enlarging the latest ring of color. In World Three the hyphae within the bracket resemble long, smooth, translucent tubes that branch and wind among each other, over and around, making a glistening matrix that cannot be unmade. The ultimate assignment of these hyphae is to reproduce: to build a body that projects from the tree, equip it with spore chutes, make the spores, and then release them.

The hyphae in the bracket all lead back to the tree, where they merge with the mycelium, a tangle of tubes invisibly buried in old cambium and heartwood. Unlike its reproductive extension, the mycelium has no particular form. Its hyphae have wandered where the adventure of absorption has taken them, an irregular, invisible mass. Its assignment is to assimilate: to extend, extend, always extend and to digest and absorb as you grow. (Keep some for yourself, but send back the rest.) What the absorbing hyphae take from the tree they pass to their brethren in the bracket to make more bracket. The grand cycle can be completed only when the spores from the polypore germinate on another tree, preferably a dead one. The spores will drift here and there in a myriad of air currents, most finally alighting on something unsuitable and then dying. But a few spores may land on just the right place, a bare and woody surface that can be digested from the start. Each will germinate, sending forth a single hypha.

The artist's conk, along with its fungal colleagues, will slowly and steadily digest the old beech. At the site of invasion, the supple, white wood is stained a dirty grey, its strength sapped. The hyphae will invade a little farther every day, insinuating themselves into the tracheids of the heartwood. The complete interconnection of the tree's inner plumbing greatly enhances the growth of the fungus, whose hyphae explore the open architecture. As they grow through sieve plates and wall valves, the hyphae secrete a powerful enzyme called lignase that dismantles the woody polymer called lignin. It does not digest the lignin utterly but rather breaks it down only far enough to expose the accompanying cellulose.

A complex of enzymes called cellulase breaks down the precious foodstuff into simple sugars such as glucose. These enter the hyphae, absorbed through its walls. Like all fungi, the artist's conk must get its carbon from sugars. Within the hyphae, they enter a sophisticated factory of further digestion, an organized soup that roils with the breaking down and building up of molecules. The hyphae are divided serially into cells, each a separate liquid cylinder defined by walls of chitin, each abutting the next through a membrane, or septum. Some of the glucose is metabolized in the cell to keep it alive. The rest is passed to the next cell, the first of many transmissions to the outer mycelium and, ultimately, to the bracket.

In Kingdom Fungi there are only three phyla, and all are present in Hungry Hollow. The names of the phyla end with the same word, *mycota,* which means "fungus," but they begin with different words that reflect the members' mode of reproduction. The spores and reproductive structures that produce them may be seen clearly in World Six.

Fungi in Phylum Zygomycota reproduce when hyphae from different individuals meet, swell, and form a *zygos,* Greek for "joining." A thick-walled zygosporangium forms. It can survive cold and dry periods but eventually germinates into a vulnerable, thin-walled sporangium that drying will kill. In a humid location, the sporangium produces spores by the hundreds. Unlike the other two phyla of fungi, zygomycetes must live in the dampest locations, in dead leaves, stumps, the sheltered spaces of soil, and water itself. Some dwell briefly in animal dung, helping to break it down.

Fungi of Phylum Ascomycota reproduce by ascospores that develop inside tubular asci, each armed with eight spores. When the time is right, the tips of the asci open to release them. Some shoot their spores with the scale velocity of a rifle bullet. Others release them gently. Many ascomycetes are pathogenic fungi that cannot wait for a plant, animal, or other fungus to die, but invade the living host directly. Other ascomycetes live on soil or wood in the forest, appearing as tiny cups, black fingers, crusts of brownish balls, or other inconspicuous forms. The fertile surfaces often appear smooth but are frequently dotted with minute chambers lined with asci.

Fungi of Phylum Basidiomycota reproduce by basidiospores produced by elongate cells called basidia. When the time is right, a basidium gently but deliberately releases its quartet of spores to fall away. Most basidiomycetes, such as mushrooms, boletes, chanterelles, polypores, and puffballs, the most highly evolved orders of the fungal kingdom, have large fruiting bodies. The basidia in most cases have no need to shoot spores because their number is legion in each body, from mushroom to polypore. Their basidia line fertile surfaces that are immensely larger than those of ascomycetes. The surface area on the gills of a large mushroom might total a thousand square centimeters. And if basidia can cover a multiplicity of walls, as in gills, they can equally line a myriad of tubes, close-packed in the underside of a bolete or polypore. Some basidiomycetes even have teeth—pores turned inside out.

Ganoderma applanatum is a basidiomycete. Its lower surface is densely and minutely perforated with pores. Each pore is lined with basidia. Down this chute the spores will rain invisibly out.

As the warm months progress in Hungry Hollow, the number of fungi in the forest steadily increases. In April and May the ascomycetes dominate, with lowly red cups and dark, inconspicuous tongues. By June the Russulas, shot with subtle pinks and yellows, spring up in wet weather. Small brown Psathyrellas hold conferences and brilliant Mycenas sport witches' caps. In July the corals appear, exotic growths seemingly imported from some Pacific reef. Scaber-stalk boletes arise from the roots of aspens, and Amanitas add a touch of deadly elegance here and there. Oysters sprout on logs, and by August, the puffballs begin to swell. Jack o'lanterns leer in glowing pumpkin colors at the base of beeches. Milkys and waxy-caps, corts and Agaricus—more than a hundred species of visible fungi ply their invisible trades. Many species appear again and again as summer fades.

If spring is the time of flowers, fall is the time of fungi. The production of spores reaches a crescendo. They float through the air of World Six, an armada of spheres ornate or smooth, of cylinders and sausages, spindles and spines. The spores will drift or blow, some a meter or so, some much farther. Every year, a tiny fraction of these (yet still in the hundreds) will float aloft, shuttled by chance winds to the upper atmosphere. Some will fall in Turkey, some in France, and some, eventually, in New Guinea.

Not far from the artist's conk, on the forest floor beneath a white oak, lives another basidiomycete. *Amanita virosa*, the destroying angel, is foresworn by Lotor and every self-respecting animal able to bite or nibble. The destroying angel has enough of the deadly toxin called amanitin to kill Lotor a hundred times over. Young mushrooms of this species have a pleasant taste. After a few bites, a raccoon might go its merry way for a day before developing cramps and vomiting. Then, like magic, the symptoms disappear and the raccoon feels better. But just when its *joie de vivre* has returned, its liver and kidneys fail, and the raccoon goes into irreversible toxic shock and dies. Equally foolish humans suffer the same fate.

Under another oak on the west side of the forest, another *Amanita* species plies its quiet trade. *Amanita muscaria* has a yellow-orange cap sprinkled with buff bits of old veil. It has little or no amanitin but harbors another drug, ibotenic acid. When he was young, Lotor once took a nibble of this mushroom and has avoided all mushrooms ever since. On that fateful day, Lotor enjoyed the mildly tangy taste for a moment, but not enough to continue eating. This was a good thing, because his metabolism had converted the ibotenic acid to muscarine. His eyes welled with tears, he began to pant, and his legs started twitching. He crawled under a bush to rest. Suddenly he was in a tree, or thought he was, but he could not remember climbing one. The leaves of the bush turned into fluttering birds, and then the entire scene gradually went out of its focus. Lotor drifted into a deep slumber.

Why do so many mushrooms live in the forest? The numbers would not seem so impressive were it not for the mushrooms that live under the great trees, specialists that arise from deep within the ground, from the very roots of the trees. For there feed the roots of these mushrooms as well. These fungi have evolved a commercial—not to say commensal—relationship with their tree partners. The intimate intertwining of hyphal roots with larger, woody ones makes a kind of market known as the mycorrhizae. The fungal hyphae explore the soil far and wide with a thoroughness impossible for the tree. Living by trade rather than decay, they bring exotic but essential minerals such as phosphorus to the mycorrhizal market. Within the cortex of the root, they exchange these minerals for sugars, the main fungal food.

Out in the meadow there are also many fungi, but fewer visible ones. Here and there a meadow mushroom sprouts after a rain and fairy rings emerge, elfen dance-floors. The great majority of the fungi prowl the soil, growing hyphae between grains, taking what is available, fruiting, and dying. Their lives sound humdrum.

Who would suspect the dark chambers of soil to be the stalking ground of a predatory fungus? The prey of *Arthrobotrys brochopaga* is the nematode, any threadworm that will fit the lasso *Arthrobotrys* carefully grows into place. It begins with the germination of a spore within the soil. A single hypha emerges from the spore, then grows into a ring of three cells, and then stops. *Arthrobotrys* waits. And waits. Most often, nothing happens at all and the organism finally dies for lack of energy. But sometimes a nematode comes slithering through the soil, his powerful body bullying aside the boulder-grains of sand. The nematode slips easily through the noose, but the instant his cuticle so much as touches the ring, it takes only a tenth of a second for all three cells to swell, contracting the ring and garroting the nematode. Cut nearly in half and unable to move, the nematode soon dies, and assimilative hyphae digest their way into the corpse. This one jackpot will finance a new generation of spores.

Consider also the zygomycete *Spirodactylon* that lives on the dung of the white-footed mouse. It does not seek the dung but is born when the dung is excreted, emerging into the fresh air that triggers its germination. It digests the remnants of cellulose, along with the sugars and other biochemicals the mouse excretes. When it finally has the energy to reproduce, the fungus grows an elaborate, spiraling tangle of spores, each with its own trailing thread.

More often than not, a mouse drops dung within its own run or burrow. Does *Spirodactylon* do maid service? In these well-frequented passages, the elaborate structure of spores suddenly makes sense. The threads of the tangle readily catch on the coat of a passing mouse and the spores follow suit, to be licked off and swallowed on the next grooming. Within the mouse's dark digestive tract, the spores resist acids and enzymes while they await evacuation. Mice do not clean up their messes, but *Spirodactylon,* with the help of other dung specialists, digests what is left. The dung will turn to innocuous earth.

The fungi live in Hungry Creek, too. There are zygomycetes that grow on plants below the water line. There are also ascomycetes and basidiomycetes that alternate between forest and stream, the amphibious fungi. In the air they release four-pronged spores that drift with the wind, landing, with luck, on the creek. In the benthos they are tetrahedral mines that drift with the bottom currents, alighting with more luck on dead leaves. Making perfect, three-point landings, they will germinate from the end of one prong.

Other phyla of protistan fungi live in World Five. Chytrids invade single cells of algae, there to flourish and nourish themselves with but one tiny hypha. Another phylum, Oomycota, features *Saprolegnia,* a giant water-mold specializing in dead insects and other animals. Taken altogether, the aquatic fungi mirror the work of their counterparts in the forest.

Let those humans who consider fungi beneath their notice consider what fungi accomplish. Let them consider a world where no one practiced the art of decay. The organic refuse would pile up year after year to the point of being no longer below notice, but above it! Fungi, along with bacteria, recycle virtually all of the annual litter of wood, leaf, bark, and corpses. The litter layer of a forest floor does not grow thicker with the years. Without the fungi to clean up, forest and field would eventually smother in ligno-cellulosic junk.

The power of the fungi goes well beyond the humble task of cleanup and recycling. Fungi may alter whole landscapes by suddenly attacking a specific plant or animal. Pathological spores may waft into Hungry Hollow on currents of air or book passage on animals. A hundred years ago, an ascomycete called *Cryphonectria parasitica* blew in to slay the three native chestnuts that lived on the ridge. Sixty years later another ascomycete, *Ophiostoma ulmi,* the Dutch elm disease, rode into the Hollow on the back of the European bark beetle, *Scolytus multistriatus.* Both local invasions happened to be part of wider campaigns. American chestnuts are practically extinct in North America, and mature American elms have become rarities.

The artist's conk was there, of course, attacking the wood so generously killed by the ascomycete invaders. For many years, bright orange and ochre brackets of *Ganoderma applanatum* decorated elms and chestnuts in the forest, eating wood, building brackets, and releasing spores.

Bear

It has taken Dianne nearly three weeks to notice the mound just inside the forest. She has walked past it almost every day. Her eyes have registered it, but the shape of the mound—its uniqueness in the otherwise flat floor of the forest—has escaped her until now. She cannot account for it, except to imagine a glacial mini-deposit, an afterthought of the ice. Or could it be an old badger-sett? (That would be something.) The mound is about eight feet long, four feet wide, and roughly elliptical.

When her curiosity can no longer be contained, she brings a trowel from her house and digs a small exploratory trench from the side of the mound toward the middle. The trench is only a few inches wide because she does not want to destroy whatever she may find in the way of old tunnels or patterns of disturbance in the soil.

The trench has progressed well into the mound when her trowel strikes something hard—not a stone, but hard. She digs around the object and pries it up from the trench. It is an animal bone, yellow with age and streaked with black. What animal and what bone? She spits on it and rubs the bone to clean it off, already beginning to suspect what it is, but afraid to think out loud. She must stay calm.

She has known all along that the mound could be a grave. That is why she went to the house for the trowel, an undercurrent of fear in her breast. The bone is from a human finger. She matches it to her hand. The name comes back from a course in anatomy: metacarpal, basal bone of finger. It is large, probably a man's.

"Stay calm. So there's a man buried here—maybe." She has begun to talk to herself in a soothing way. The trowel, with a life of its own, moves back into the trench and exhumes several more bones, all belonging to someone's hand. The bones have a direction. The radius and ulna must be over here on the left. She pushes the trowel in, hard, until it strikes more bone. "Yes, okay." She says gently. She is sweating. "And in the other direction we should find nothing."

But there is something there, not a bone but a root. Not a root but something softer. It gives under the force of the trowel. She digs around the object and then brings it up to the daylight. It is covered with dirt. She picks away the particles of clay and silt until she sees that it is a small bag. Guiltily, she looks around and then puts the bag in her backpack, hastily shoveling the earth back into the trench, tramping on it, patting it smooth, and covering it with leaf-litter again. She stands up to inspect her handiwork. Can anyone tell that someone has been digging here? No way.

At home, she carefully scrapes the bag with a knife to get the dirt off. It is made from disintegrating hide and threatens to fall apart. She can feel things inside it. She knows that it is a native grave now, and she shakes with excitement. Before opening the bag she stops to have a mug of coffee, but that only makes her more nervous. It is time to open the bag.

She lays it on a piece of newspaper and pulls gently at the rawhide strand that ties it. The string falls apart in her hand and the bag is suddenly open. Gingerly, she draws the contents forth into the daylight of her kitchen: There is a skin, faintly rosy, with fine down falling away from it. There is a long bone, pointed at one end and knobbed at the other. There is a set of claws—bear claws she thinks—ten in all. Next there is a seashell, like a small conch. Finally there is a dark blue stone with a perfect white ring around it. Nothing in her training has prepared her for this moment, but the deteriorating bag is, she thinks, a medicine bag.

What will she do? She puts the bag in an old cookie tin and adds a moth ball to keep insects from eating what is left. She puts the cookie tin in the cupboard. She will not go into the Hollow again today. She is left with the dilemma of what to do. She reasons, not without guilt, that because the body has lain in its shallow grave for who knows how many hundreds of years, it won't make any difference if she says nothing for a few days. Or weeks, for that matter.

She will never know that the body dates back to the summer of 1591. A band of Shawnee had lived in the Hollow for five years. They had burned down several large trees at the edge of the flood plain and made some crops of beans and corn. At that time the forest came closer to the creek, but there was still room for bark lodges. Children played along the shore, the women cut down a black ash tree to make baskets, and the men hunted deer and wild turkey.

One day their shaman, Wassamowin (Lightning Tree) announced that his totem would pay them a visit that evening.

This made everyone edgy because their shaman belonged to the bear clan. As they went about their business, they glanced nervously around, wondering when the bear would show up.

That evening, just as the sun touched the horizon, a bear came out of the forest above the stony cliffs and looked down at the bark houses. It was not *a* bear, but Bear. The power of Wassamowin was awesome. His power came from Bear.

When he heard the silence, Wassamowin came out of his lodge and sang a short song. His voice was high and penetrating. It seemed to fill the hollow. Then Bear reared up on its hind legs and made a strange noise. Beginning as a squeal, it descended into a throaty rumble.

The people trembled with fear, each rooted to the spot. Wassamowin spoke to Bear for all to hear. He asked for favors: plentiful game, good crops, success in battle, and good health for the newborn.

Bear roared again, just as it had done before; then it went down on all fours, turned, and walked back into the forest. For the rest of the summer the people felt the eyes of Bear upon them. The feeling grew so eerie that they knew they would have to move soon. They called the hollow Makwakijadon, or "Bear Watches."

One fall day there came a boy of twelve summers to the lodge of Wassamowin. It was his spirit time. He sat trembling outside the door until the sun was past its height, fearing even to look inside. Then Wassamowin appeared suddenly at the door and stared at the boy with a face so contorted, so fearsome and wild, that the youngster jumped in fright. Wassamowin laughed until he could stand it no more.

He brought the boy inside and put some magic herbs on the fire until the lodge was heavy with perfumed smoke. He said:

"They called the hollow Makwakijadon, or 'Bear Watches'."

"The time of change is upon you. You have waited for this day with fear and hope, which is only proper. Whatever path you walk, fear and hope will be your companions and helpers. Watch the fear, that it helps by making you afraid at the right times. Sometimes a fear begins to think it is real, like a person. So it tries to become the person it belongs to. Watch that your fear does not think it is you and try to take over your life.

"And watch your hope, that it does not become sick so that your heart grows heavy and disease comes. If your hope should die, you will be in an awful fix! For you will heave a great sigh of sorrow and hopelessness and cry like a baby, even though you are a man.

"The time of the vision is the most important time of your life. Everything you do after this will be different from what you did before.

"Go to the forest beyond the ridge and choose a strong tree with two strong branches close together, like this. Take rawhide and a robe. Make a bed in that tree and stay there, fasting. As the days go by, you will see and hear many strange things. Each time something happens, sing this song to test it."

Wassamowin sang the song to him.

"If the apparition is from the Trickster, it will disappear when you sing. But if it comes from the Creator, it will not disappear but grow stronger. This will be your vision. Remember it well. When you return, I will announce your name."

The boy did as the shaman bade him. He made a hammock in a large tree and stayed there, day and night, until his stomach knotted with the pangs of hunger and his head throbbed. He wanted to go home, but if he tried to sneak back, Wassamowin would know about it, somehow. His home seemed infinitely sweet now. He pictured his mother and father, his brothers and sisters, sitting around the evening fire to eat roast quail and corn soup.

After this, the boy grew faint and could barely move. Then he saw raccoons dancing in the air all about him. It was so strange and amusing that he almost forgot to sing the magic song. When he did, the raccoons all disappeared.

Then he heard the sound of a bear and said to himself, "That is Bear." Just to be sure, he sang the song again, but Bear heard him and only roared louder. It was coming to his tree. He could feel the forest shake with Bear's footsteps. Then he heard a huge scratching sound and felt the tree tremble as Bear stretched for the boy's hammock. It could not quite reach that high, so it dragged its claws down the bark of the tree instead. Then it looked intently into the forest, smelled the air, and took off, as though afraid of something. The boy wondered what Bear could be afraid of.

By this time the east was growing red with the rising sun. Even as the red grew brighter, the boy began to hear distant voices that came from where Bear had looked. As the voices came nearer, the boy could hear that they were not speaking Shawnee. The strangers gathered beneath the vision tree, and the boy grew so curious that he looked down from his platform.

Their skin was as white as birch bark, and hair grew on their faces. Some of them had white eyes. They argued and carried sticks with which they beat everything around them, bushes and stones. The boy could see that they were insane, and a great fear came over him, along with an unaccountable sadness that was never to leave.

When the sun came up the strangers all disappeared, and their voices faded to nothing. Suddenly the boy's strength returned, and he no longer felt tired and weak. He forgot about his starved condition and climbed down from the tree. His one thought was to go immediately to Wassamowin. The vision was over.

When he arrived at the lodge, Wassamowin looked up from the coals of his fire.

"Who are you?"

"I am the boy you sent to the forest."

"Oh, I don't think so. I've never seen you before in my life. But you look hungry. I never turn away a hungry stranger. Have some of this soup and take it slowly. Tell me how things are in the forest, but don't rush."

The boy told him about the dancing raccoons and Wassamowin laughed. But when he told him about Bear and the vision of the white-faced strangers, the shaman's mood changed suddenly, and he stared fixedly into the fire. Abruptly, he got up and left. When he returned, some of the elders accompanied him. They sat in a sacred circle.

"Today a stranger has come into our camp and I think we should adopt him. His name is "Wabishkingwe.""

The name meant "White Faces Man" and the elders wondered how he got it.

"I have not spoken of this before, but now I must. It is a sacred matter not to be yacked about over the fire. It is this. In the time of the Prophet, during the pawpaw moon, there came a vision to him of hairy-faced men whose skins are pale. They live somewhere over the water and their medicine is fearsome."

Wassamowin reached behind him and brought forth a small deerskin bag that everyone recognized. They grew fearful as Wassamowin shook the bag in front of them. "Even their children have more power than this." He was silent.

"At first the Prophet was glad, for the medicine was enormous and would be a big help to everyone. But there came a second vision to the Prophet at the end of the moon. He was taken across the water in a dream by the Great Eagle and moved among the hairy-faces. Through the all-seeing eye of Eagle, the Prophet could look right inside the hairy-faces. He could see their hearts. Their hearts were shriveled up and black, like rotting plums. And some of them had no hearts at all. It was Trickster who did this, and this is the manner of it.

"Trickster gave some of the hairy-faces a little of his medicine, and they liked it so well that they wanted more. So Trickster said, 'Give me a bit of your hearts and I will give you more medicine.' At first the pale people didn't want to do this, but some of them went ahead anyway. As soon as the others saw what Trickster was giving out, they wanted some too, so they also gave up some of their hearts. Before long, Trickster had given them all his medicine, but he had their hearts, which were far more valuable. He kept them in a huge otter-skin bag, and only Trickster knows where the bag is hidden.

"The hairy people did not mind having no hearts, for they were too busy trying out their new medicine on all kinds of things. They enjoyed it.

"Wabishkingwe is the only one of us to have seen these people. It is a great vision, but a terrible one.

"The hairy-faces are coming here. Some of you will hear of them before long, and some of you will see them for yourselves if you live long enough. Aiieee! What tragedy they will bring! Death and destruction of everything is their aim, for they are already dead and will use their powerful medicine to make everything like them. I thank the Creator that I will be leaving soon. Aiieee!"

With that the assembly ended and everyone returned to their lodges. The new man Wabishkingwe returned to his lodge.

"My name is Wabishkingwe. May I stay in your place for a while and live with you?"

"Certainly," said his mother. "I have some quail for you and corn soup."

Wassamowin died that fall of a strange ailment. They buried him in a secret place in the forest, along with his medicine bag.

Stories in Stone

Who dropped the granite boulder in the upland forest? It rests, buried to its waist, between two oaks. Angular still, with barely rounded shoulders, it remembers time. It was brought to Hungry Hollow in the frozen heart of a glacier, but it was born long before in heat so intense that rock ran like water.

The water of Hungry Creek runs over rocks that are smooth and round. They did not come from far away, nor are they so old.

The story of the stones is written in a script called Old Atomic. The letters of its alphabet are aluminum, boron, carbon, and on to zirconium. The words are crystals of quartz, fossils of carbonate, flecks of gold, flakes of shale, drops of oil, cobbles of coal, all deriving directly or indirectly from the molten alphabet soup called magma. The text is piled neatly in some places, tattered in others, and consumed elsewhere by the inferno. The great heat is not far from Hungry Hollow. It is closer than the big city, straight down.

The larger story begins at the zigzag zones where new crust forms and ends with its own subduction into magma. Its plot is elaborated by global and local forces, by weather and biology, by time and by space, into a myriad of subplots, stories within stories. There are at least two entries for Hungry Hollow:

Limestone

Now and then, pieces of the limestone ledge above the bend of Hungry Creek break free and fall into the water. The legacy of this ledge runs downstream for many kilometers. At the bend itself, several chunks of limestone, rough and jagged, lie just below the surface of the creek. They refuse to budge, even in the powerful currents of spring. But they are wearing down, and next year one of them will tilt and roll over. One small step.

Downstream, the stones grade from jagged to smooth, getting smaller. If days were seconds, the stones would move slowly at first, then tumble, then roll downstream in a rush to the mouth of Hungry Creek on the Askunissippi River.

At the bend itself there are many large, round rocks that did not come from the ledge above. Some originated upstream from the same formation, but others were smooth long before Hungry Creek began to flow. They lay buried in the Hollow until the creek, in its restless meanderings, excavated them. These early stones rolled down another river a hundred thousand years ago.

The story of the limestone begins millions of years before that. The ledges of Hungry Creek were born late in the mid-Devonian period, almost four hundred million years ago.

"The legacy of this ledge runs downstream for many kilometers."

Great shelves, running for tens—even hundreds—of kilometers, were built by the collective secretions and body parts of fish, trilobites, molluscs, lamp shells, corals, bryozoans, sea-lilies, sponges, and algae.

The great inland seas that gave them birth slowly closed up and drained away as the North American plate broke loose from an early Europe, drifted for fifty million years, and then closed again with other continents to make Pangea. The supercontinental patchwork held together for a hundred million years, then slowly broke apart again. North America began its grand swing into the northern hemisphere, joining hands with South America and rafting gradually to its present position.

All this time, much of the shelf that formed in the inland sea was buried ever more deeply in sediments that accumulated like the pages of a book, turning, age upon age. Pressure and heat increased steadily. The carbonate minerals from shells, corals, and crinoids began to creep through the old shelf, permeating softer bodies of protein and chitin and everywhere crystallizing minutely to calcite. The traces of life fossilized, locked in an ever-hardening matrix of limestone.

The book of sediments, whose pages are layers, became worn and tattered. Whole chapters eroded away during continental uplift, washed by ancient rivers and ground down by the monstrous glaciers of the Permian and Carboniferous periods and of the recent Cenozoic. The latest ice age is still with us. We live in a warm but brief interglacial spell, the one that has nurtured all of recorded human history.

The local page is the limestone ledge. Deeply buried until ten thousand years ago, its ancient lime was destined to surface once more in the story of life. It dissolves again as stones from the ledge erode. Clams and snails eagerly absorb the carbonate molecules, some of which once belonged to much, much older clams and snails. Yet the letters of calcium, carbon, and oxygen are as new as ever.

Granite

Fifteen thousand years ago, a huge lobe of ice bore down on Hungry Hollow, a sheet hundreds of meters thick that oppressed the land in grinding darkness. It moved slowly across the face of time, the hour hand of a cosmic clock. Above the glacier, the sun shone through haze and cloud on snow and ice in all directions—a frozen plateau that spread to the northern, eastern, and western horizons. To the south it fell away to frigid plains of tundra and boreal forest of spruce and tamarack, where now the sweetgum and magnolia grow.

Because it moved and because it was so big, the great glacier was denied nothing in its course. A south-facing granite cliff far to the north had endured for millennia. Encased by the dark pressure of a kilometer of ice, it split slowly in the heavy silence, giving up 240 tonnes of its face to the glacier. Pieces of the plucked cliff fell out thousands of years later when the glacier began to melt. But the ice age deepened again. And again the glacier swelled, and again went on the move, carrying far below its snow-swept surface a granite block destined for Hungry Hollow.

The boulder had sharp edges at first. Set like a fang in the jaws of the glacier, it gnawed on granite ground below, leaving long tooth marks but gradually losing its bite. At the end, when the climate warmed once again and the base of the glacier melted back, the boulder emerged into sunshine a rounded thing. The old glacial erratic now sits in the forest of Hungry

Hollow, knowing what we do not, waiting with geological patience for the current ice age to resume.

The great cliff from which the boulder split had a more ancient tale to tell. Long before it became a cliff, it lay fused within a great granite formation that was thicker than any glacier. The first and ancient continental crust of North America formed four billion years earlier when Earth was still young. Up-welling magma slowly cooled as it rose. Liquid rock turned solid, crystallizing into grains of quartz, mica, and feldspar, locked at last into durable granite.

The surface of the proto-continent, washed by the first rains, cooled into hard crust. The granite destined for the future cliff still lay deeply buried.

It would take billions of years for the same rains to wear the granite land away. The atmosphere of carbon dioxide and nitrogen enhanced the wear, dissolving feldspar into clays. Within the feldspar pits, it attacked the mica and other, more resistant minerals, leaving in the end a crumbling surface of quartz crystals. Sand. These grains washed away to form sandy beaches and shores along rivers and coasts, though there was no one to stroll on them.

Nothing was visibly alive, yet some waters had a strange discoloration. By the time the top of the future cliff appeared, the discoloration had turned into primitive plants and animals. By the time the cliff was plucked by the glacier, most of the living things of Hungry Hollow had already evolved.

The boulder waits in the forest. The next glacier will wear it down, and what remains will weather to clay and sandy sediments. Some day North America will collide again with another continent, and whatever remains of the boulder will slide imperceptibly, beneath the collision, into the abyss of magma.

It will melt in the inferno. The last remnants will dissolve back into the most common letters: oxygen, silicon, aluminum, iron, and sulfur.

The story will end even as it begins anew in the mid-Atlantic and in the African rift valley. Wherever crust breaks apart, magma wells up to make new words and sentences.

Hungry Creek

You cannot tell what minerals compose the stones of Hungry Creek unless you pick them up and turn them over. Their tops are coated with an olive biofilm, and they trail long green tresses in the current. The creek enters the Hollow though a two-meter culvert under the highway, and here the stones are particularly large and plentiful. The creek spills over some of them, cooling and finding relief in the shade of willows, aspens, and a lone sycamore. It has traveled through fields of corn and soybeans to the north, overheated in the sun and overloaded with nitrates and phosphates from fertilizer.

Algae that bloomed upstream have primed the food web of Worlds Four and Five. Green mandalas of *Pediastrum,* translucent rafts of *Scenedesmus,* and long tubes of *Spirogyra* and *Cladophora* drift amid a scattering of diatoms and cyanobacteria. Creations of the sun, they are the excess production of the upstream creek. Humble flagellates and bacteria accompany the caravan, hunted by cruising *Paramecium,* by the endlessly spinning *Urocentrum* (of the species *turbo!*) and by the agile *Cyclidium* that flicks from place to place faster than the eye of any world can follow. *Coleps* and other predators hunt bigger game. All are immigrants, recruited from the bottom into the driving current by the excess of their numbers. Streams have no obligate plankton.

Schools of dace and minnows hover in the shade, aerating their gills in the current; sticklebacks and darters prowl the stony bed, grazing on the biofilm, a source one hundred times as rich as the open water.

Summer is fading. A single leaf, coated with road dust, detaches itself from the sycamore and sideslips through the early September heat. It falls into Hungry Creek and begins an hour-long journey through the Hollow. It has landed near the shore of the creek and moves slowly along the bank past mossy gardens, rainforests of World Three. The leaf drifts toward the center of the stream where it picks up speed in the faster current. It spins for a moment, caught in an eddy that wells up in a complex dynamic of its own.

Velocity profiles govern the flow of Hungry Creek. Beside the bank the current is sluggish, but toward the center a horizontal transition hurries it to a walking pace. Water striders, the long-legged bugs that skate on surface tension, juggle the currents to their advantage like hawks riding thermals. They skate near the shore to save energy. But they stride upstream against the central current when they are well fed by what it brings in the way of small floating insects, dead or struggling.

There is a second profile that extends vertically from the top of the stream to the bottom. A modest meter per second at the surface slows with depth to a centimeter per second at the very tops of stones where the larvae of caddis flies and midges

"As it swings around the bend, the sycamore leaf is caught in a roiling cross current."

must live in anchored cases. Between the stones the current dwindles to almost nothing. Yet in the scale of World Five, such speeds are all more or less enormous. The surface current of 100,000 meters per second slows with depth: 10,000, 1000, 100 meters per second. Next to the basic benthos, a tenth of a millimeter above it, the boundary layer of creeping water flows at a scale speed that is still tricky for protists to negotiate: Ten meters per second will pluck away the slow ones, yet *Paramecium* and the chubby, voracious *Pleuronema* may power along at twenty meters per second in World Five—enough to swim upstream, provided they stay low.

The larger creatures of Worlds Four, Three, and Two have a slightly easier time of it. Ostracods and water fleas can venture up a few millimeters from the bottom, amphipods even higher. Larvae of dragonflies and damsel flies make their way stealthily between rocks, hunting small crustacea and aquatic insects. Molluscs have nothing to fear in the current and pursue private lives in complete absorption. Clams are too heavy for the current to budge as they inhale tiny immigrants. Snails keep a death grip on the substrate, grazing on the biofilm.

Closer to the shore, whirligig beetles make frenetic circles, and water spiders skitter crazily out from shore to seize a struggling insect and then beat their way home. Below the surface, backswimmers and bug-eyed water boatmen oar speedily through the water, while diving beetles hunt other insects or their larvae. Crayfish lurk beneath rocks, awaiting the night and a chance to forage for rotting vegetation by the bank.

The sycamore leaf follows the creek along a wide curve to the left, still in the shade. The creek approaches a cliff where it has worked for many years excavating the bank, mining for fossils. At the cliff, the creek bends sharply and the profile of speeds shifts toward the bank. The fastest current is thrown here by centrifugal force. It has the power to gouge and lift away bit by bit, layer after layer. Small grey plumes darken the water as the shale returns to active duty after 380 million years of shore leave. Across the creek from the carved bank, the current slows to nothing over shallow water.

As it swings around the bend, the sycamore leaf is caught in a roiling cross-current that takes it to the slow shore and leaves it spinning slowly in an eddy. Here the creek meets an older version of its bed. The resulting shallows make a small bay that intrudes inland several meters. It is planted with spatterdock, Elodea, and coontail; embroidered with duckweed; and framed by arrowhead, cattails, and bur-reeds. Behind these, jewel weed, marsh ferns, and Joe-Pye weed have established colonies in the wet soil. Here and elsewhere in the wet margin lands of World Two, pickerel frogs, leopard frogs, green frogs, and a lone bullfrog have their base of operations.

The shallows play host to lovers of still water. Seed shrimps, crayfish, and amphipods scavenge the bottom for organic wreckage. They scramble or swim on jointed legs, nimble crustacean arthropods in three convenient sizes for predators. Minnows, soft-shell turtles, and Lotor himself will feed on them. Lotor will also seize a small painted turtle that has seized a minnow. No one is safe in the shallows. Some of the larger fish (chub, carp, and sucker) sometimes visit the shallows to vacuum the bottom, but the game fish (pickerel and trout) disappeared from Hungry Creek decades ago, fished to extirpation or poisoned. At the moment, the shallows are not for fish, anyway. Two minnows float belly up near the surface.

The leaf is carried by a slow current clockwise around the mouth of the shallows. At the end of the shallows, a pair of human hands dip a sample bottle into a discolored, rank-smelling patch of water. It is Dianne, who has noticed the algal bloom. She stares at the contents (a blue-green discoloration), sniffs the bottle, makes a face, and then snaps on the lid. She will discover that a cyanobacterium called *Microcystis* has responded to excess nutrients from the upstream farmlands. It has reproduced exponentially, making the water in the shallows a thin, blue-green soup and releasing toxins that paralyze fish.

The leaf resumes its journey at the whim of Hungry Creek. It departs the shade of leaves and, washed of its dust, enters the full sun again in the meadowlands. The tiny factories of photosynthesis within the leaf are prompted to produce once more, even as they die. The leaf ripples ironically over the stony bottom fifteen centimeters below. It passes carp and suckers, dace and darters. A belted kingfisher, perching in a wild plum, eyes the leaf suspiciously as it passes beneath his branch.

Hungry Creek sweeps gradually to the center of Hungry Hollow, where it meanders slightly within moderate banks overgrown by reeds and grasses that lean out over the shore, shading the slow-moving edge-water. It is late afternoon on the floodplain meadow.

The sycamore leaf halts abruptly, snagged on a submerged branch.

Wonder of wonders! It has been more than a year since a deer last visited Hungry Hollow. Its leaf on the Tree of Life fell some months ago, but now it is back. The white-tailed deer, *Odocoileus virginianus,* has returned in the form of a yearling buck whose mother has been run off or run over. Alone now, with little education, it must puzzle out its existence from the book of leaves. Some taste good and some not. It pursues the knowledge of good plants and evil from woodlot to woodlot, the possibilities delimited by a grid of highways and roads that enclose occasional miniature forests. If it must watch for cars now, at least it has no cougar or wolves to fear. The leaves of these burly predators fell from the Tree of Life more than a hundred and fifty years ago, when roads first dissected the country hereabouts.

Dianne trudges up the cliff path to her home but fails to look over the meadow. She does not see the deer pause by the edge of the creek, bending its small antlers to the water.

Without fear, the deer drinks a long draught from the cool waters. Little he cares for the animated cocktail within. The taste seems clear, refreshing. He raises his head again, sniffing the air for faint scents of his favorites: spice bush, alder, sweet grasses and sedges. A sudden whiff of suburban gas wrinkles his nose. Fear seizes him. He bounds across the creek in three splashes, one of which dislodges the leaf from its snag.

Hungry Creek meanders first right and then left as it nears the end of Hungry Hollow. In the upland forest, cicadas buzz with the joy of heat. The creek enters a cornfield and makes a sharp turn into a channel dug by backhoe. A crow, perched on the nearby fence, sees great significance in the leaf and caws in alarm. It flaps up from the fence, into the humid heat. It looks down on Hungry Hollow and scans the sinuous course of Hungry Creek.

A million crows have done this over time. They have inspected the Hollow from every angle and have seen the creek in every position, from one side of the floodplain meadow to the other. There is no part of the floodplain that the creek has not crossed. The mechanism of this continual change of course is simple. At each bend, however modest, the outer current is stronger than the inner. It cuts new outside banks, drawing soil into the water. The inner current slows enough to drop its load of sediment, growing a point bar. Bends get wider, curves more exaggerated. Loops form, then cut off into oxbows that dry into chains of ephemeral ponds that fill within a hundred years. Speeded up in time-lapse photography, the creek would writhe like a snake, cutting and filling.

It has excavated the valley down to its present level and will continue its work into the predictable future, always taking soil out and bringing in slightly less.

In the summer Hungry Creek moves slowly, carrying only silt and clay. But in early spring it floods with powerful currents fed by melting snow and ice. It tumbles stones along the bed for meters at a stretch. It runs with sediment, like rich coffee. It sometimes floods the meadow with a shallow, temporary lake that drops its suspended load in fine layers, rich with nutrients for plants.

Three small exceptions barely mar the creek's monopoly on the transport of minerals. There is first a daily rain of meteoric dust too fine to see, sediment from space itself. Second, tiny grains of sand from the Sahara, the Ogaden, and the Rub' al Khali filter down from stratospheric winds. Third, there is another granite boulder in the Hollow besides the one in the forest. This one lies nearly buried by the east bank of the creek. It has moved barely a meter downvalley since it melted out of a glacier ten thousand years ago, but it has sunk five meters during the gradual excavation.

The crow spies a small oddity among the goldenrods near the edge of the forest. It caws out and swoops down for a closer look. Yes. Yes. It is something dead! A toad. With a rustle of ebony wings, the crow lands beside the corpse and cocks its head the better to see. It struts stiffly back and forth, then takes an exploratory peck. No response. Other crows come. They must finish their business before the turkey vulture spots the prize.

Requiem for a Toad

He lies on his back, belly wrinkled and hard. Ants and flies cling to the carcass as the crows pick and peck. He was a male American toad, the most common amphibian in North America. The species called *Bufo americanus* is not in danger, but he was. Bad luck. At the age of three years, he had the misfortune to enter a statistical trough, a natural event for all species in all habitats.

An adept hunter, he nevertheless encountered a pocket of sparsity. There were no catchable insects for a period just long enough to weaken him seriously. Even when he felt incapable of crawling another centimeter, nothing came, and when the flies finally and insolently landed on his nose, he was barely aware of them. His superb tongue lay swollen and inert in his throat.

When the crows had picked him clean, the ants resumed their foraging unhindered and the flies returned to spit and suck on bits of meat too small for beaks.

"Our dear departed here was at first a black speck within a clear, round sphere."

The First Crow Speaks

"My friends, let us pause to consider the toad whose remains lie here before us. It is meet and fitting to give thanks for his existence and to recount his story.

"He was born three springs ago in the shallows of Hungry Creek. The winter had been long and the breeding season came late that year. Thirty-two males showed up for the party. They sang long and loud, filling the hollow with their trills. They sat in shallow water near the shore, inflating their throats to speckled white balloons that magnified their calls. Five females joined the party, by and by. Our dinner's mother was dinner enough for ten, a large lady who was seized by a handsome and aggressive young fellow waiting at just the right spot. Good luck. His bagpipes impressed this grand dame of the upper hollow.

"Other males tried to rip him from the back of his intended. Sometimes two would try, and the four toads made a knot that sank in the shallows until you'd have thought they would drown. Other males, with no partners, made small, horrid, burbling noises and rushed at each other to rip out intestines with sharp claws. Some were mistakenly mounted and gave the release call. Grr-umph.

"The grip of the male toad is fierce, unbreakable. After a night and a day, our dinner's mother finally shuddered and began to spew out a string of eggs. Her lover's reckless expenditure of sperm coated the eggs. The mother swam clumsily around the spatterdock, winding the eggs to stems. Her lover rode her, prisoner of his own grip. Not until she surfaced and made for the shore did he realize it was over. He let go. It had been an exhausting copulation. He was hungry.

171

"The eggs, thousands of them, were all alike. Our dear departed here was at first a black speck within a clear round sphere, set in a mass of jelly. In a week a noticeable tail had developed, and the black form within began to wriggle every few seconds.

"The tadpole broke from its egg but remained attached to the gel, sucking on it, swelling on its food. Then, a day later, it wriggled at random through the warm waters of the shallows. Although tadpoles taste unpleasant to many, the killdeer rather likes them. Every day she ate her fill of tadpoles that strayed too close to shore. Luck was with the deceased. He swam farther out.

"The young tadpole lived in World Three and played the gourmand on the creatures of Worlds Four and Five. He swam to eat, like a baleen whale. With mouth agape, he took in everything that came his way, algae and ciliates both. He spent less energy swimming than he took in as food. His gut grew apace. Then he felt increasing attraction for the shore and rooted himself to one spot, vacuuming. The killdeer came near him twice.

"After several weeks of this life, two legs sprouted beside his tail. Then two tiny arms grew from his chest. His tail began to shrink and he felt an urge to crawl. The day came when he wriggled out of the meniscus and broke free into the sun. His skin dried for the first time and he hopped, somewhat clumsily, into the grasses by the shallows. He and a hundred other toadlets passed a garter snake that day. It slithered calmly back and forth along the shore, systematically swallowing every baby it could find.

"What luck! While the other young toads met with various mishaps—eaten by snakes, birds, large toads, skunks, and other creatures—our friend, not one whit superior to the others, survived. Within two months of hatching, the fifteen thousand tadpoles had been whittled down to a thousand peep-toads. They earned their keep within a patch of territory that slowly expanded until it covered the entire valley west of Hungry Creek. By fall, when the young toads had grown to two or more centimeters in length, there were only two hundred left.

"His tongue was the tool of his trade. He would sit by the base of a curly dock, watching a ladybug descend the stem. Ka-zong! The tongue would dart out like lightning and stick firmly to the ladybug, yanking her back into his mouth. He never missed. He ate well.

"One cold fall day, our dear departed found a patch of leaves at the edge of the forest and dug himself slowly under them into thin, humic soil. He had grown cold and could barely move. Winter brought snow and sleet. His temperature dropped almost to freezing. He hibernated, sleeping the sleep of the near-dead.

"The following spring, when the ground warmed, he emerged famished from the litter to find few insects about. He survived on snow fleas and early mites. With the coming of warm weather, he found food more plentiful, graduating to mosquitoes, flies, and small ground beetles.

"Although many toads from previous spawnings inhabited our valley, the fates of the yearlings that year make clear the odds. Of the 125 that emerged from hibernation, having survived the cold, ten were eaten by Lotor and that other raccoon, whatsername, that lives north of the road.

"Have you watched a raccoon eat a toad? I watched Lotor once from a tree. To avoid the poison on the head, he flipped it over and tore at the skin of its belly with sharp nails. He disembowelled it, gobbled up the guts, and began to chew on the legs and arms. When he finished, there was a sort of shell left (still quite filling, I assure you). Where was I?

"Our fellow birds (Class Aves) ate another sixteen young toads, skunks got five more, and six died of natural causes, including disease and bad luck in choice of territory. Starvation, I mean. That left 88 toads to carry on.

"Our friend, who lived all this time in the forest, had grown to five centimeters, nearly adult size. He survived in part because his coloring changed to a very dark brown with black warts. (Black is such a safe color.) As long as he didn't move, you wouldn't see him in a month of Sundays.

"He spent his second winter in the forest, like the first, and emerged successfully in the spring once more. His class of the two-year-olds had been whittled down to 53 by now. When the mating party began that April, he was one of the first to rush to the shallows to sing. But his inexperience led him to plunge too far out; he had no place to sit and conserve his energy like the more experienced males. Instead he rushed about, scratching at the others and burbling fearsomely. He made quite an impression as an up-and-comer, but he got no mate. In desperation he jumped on a male leopard frog, thinking it a small female. There was terrible confusion.

"In his third summer, this very year, he moved from the forest back to the meadow, changing his color to cinnamon and wandering far in search of insects, spiders, millipedes, slugs, and grubs. All graduating classes considered, the toad population in the meadow peaked this year. Competition was fierce. The July heat wave left many a plant wilted and many an insect to wander to greener parts. Our deceased subject became somewhat stranded. His meals grew few and far between, weakening him to the point where he could no longer consider a serious migration. The rest of the story you know."

At this the First Crow dropped his head and stared mournfully, but in a satisfied way, at the skeleton before them.

"Poor fellow," said the others.

"I beg to differ! He was among the luckiest of the lot. At his demise, there were only fifteen of his generation left in Hungry Hollow. Let us not mourn but be joyful on his behalf."

One of the crows mentioned that he had seen some children in the nearby suburb carrying food around. With caws of enthusiasm, the murder flapped into the air as one and flew raggedly to the line of trees on the bluff to spy on Whispering Pines.

No fewer than 26 species of insects, six species of fungi, three species of protists, and eight kinds of bacteria took part in engineering the disappearance and recycling of the corpse. Within a week the toad had vanished into thin air.

Survival
of the Lucky

Within every cell of an organism, within every nucleus, lengthy strands of DNA wind their way through World Eight, where atoms are balls and the smaller molecules will just fit in your hand. In this world, you might pick up a strand of DNA, a spiral, ropy ladder made of large beads, a complex rosary on which prayers for survival may be said.

The rungs of the ladder are made of hexagonal molecules called bases: adenine (A), cytosine (C), guanine (G), and thymine (T). Each rung consists of two bases linked together in just two possible combinations: T–A and C–G. The sides of the ladder are serially linked sugar–phosphate groups, each group attaching to a rung on its side of the ladder.

In World Eight, a single, spherical coccus bacterium might be 100 meters wide and its DNA at least 150 kilometers long. The DNA must therefore be wound up inside the bacterium, bunched and doubled over many times. In the larger cells of eucaryotic organisms, from fungi to animals, the DNA is packed into a nucleus that is bigger than most bacteria. In World Eight, the nucleus of a human cell might be half a kilometer wide, containing enough DNA to stretch 100,000 kilometers.

The American toad, *Bufo americanus,* is a diploid organism, as are most animals. The chromosomes of every cell contain not one but two identical strands of DNA. During the breeding season, however, male and female toads produce gametes, reproductive cells that are haploid, with one set of DNA strands each. The haploid nuclei fuse when sperm meets egg, and the foundation is laid for a new organism that inherits DNA from both parents, diploid again.

The DNA of a cell directs its growth, maintenance, and behavior. Thus in the developing toad embryo, the initial cell will divide again and again, each time duplicating its paired strands of DNA. In the process the helix unwinds and the base-pairs split apart. The ladder breaks into two halves, and the cellular machinery supplies new bases and new sugar–phosphate groups by which each of the half-strands may be completed. In one of the half-strands, a sequences of bases such as

T T A G C G A

can be completed in the new half-strand only by the complementary sequence:

A A T C G C T

The new double helix thus matches the old.

Directly or indirectly, the DNA of a cell controls everything that goes on, from metabolism to division. It controls the process by which cells become specialized within larger bodies and how they ultimately contribute to their specialty, whether as kidney or nerve, root or leaf. Single-celled organisms behave according to the dictates of their DNA, the closest thing they have to a brain.

A new cell builds itself from protein and other materials under the direction of DNA Central. The nucleus communicates with the rest of the cell with longish molecules of messenger RNA, strands that are very similar to DNA but of limited length and lifetime. Within the nucleus, a short portion of its DNA may be transcribed into messenger RNA that drifts, impelled by the random dance of heat and cellular currents, into the surrounding cytoplasm. The word has gone out: "Make such-and-such a protein."

When the messenger RNA encounters one of the many ribosomes, protein factories of the cell, the ribosome literally reads the message, processing it sequentially, three base-pairs at a time. This code is ample enough to specify which of twenty amino acids to produce next. The ribosome bonds each new amino acid to the last, and as the new protein emerges from the ribosome, a curious thing happens. The chain of amino acids begins to twist. It writhes and quivers as electronic imbalances between different portions of the emerging molecule attract or repel each other. A three-dimensional shape comes into being, the translation of pure information into functional form, the word made flesh.

The new molecule might be an enzyme to catalyze digestive processes, to break molecules down or build them up. It might be a machine part: a cellular wall, microtubule, ion pump, or flagellum motor. It might be a protective protein: collagen of skin, keratin of feather, chitin of fungal walls or insect armor. The balance of cellular activity is driven by the proteins made on ribosomes, and, because the individual is determined in this way, its success depends on its DNA. Almost.

The ultimate fate of an organism is only partly determined by its DNA, the genotype. Its destiny also arises from the influences exerted on it by the environment itself. Here the mass of cells that make up the organism—the phenotype—must survive. Its success in the enterprise will depend in part on how well its structure and function match the environment. A toad that cannot hop or shoot its tongue is in trouble. It will die. But a toad that does not hop so well as another or shoot its tongue so accurately may still survive to reproduce.

Genes are (relatively) short stretches of DNA that control the development of specific characteristics in an organism. The DNA that determines a species such as *Bufo americanus* is not precisely the same in all individuals. For example, many genes may occur in different forms, each form encoding a different variation in the characteristic it controls. When the single strands of parental DNA join to produce the diploid tadpole, for example, the offspring might inherit different versions of the same gene from the two parents. Sometimes one gene dominates the other; sometimes the characters so encoded blend.

Now and then, parental DNA in the cells that produce gametes undergoes other changes. Corresponding stretches of the paired strands of DNA may cross over, for example, one stretch replacing the other. This will produce a different genetic inheritance for offspring.

Occasionally parental, germ cell DNA is struck by a cosmic ray from outer space, a fast neutron from the radioactive interior of the earth, or a hard ultraviolet ray from the sun. Or it may be disrupted by an environmental toxin. The DNA is repaired by the parent cell in all cases, but sometimes the information is lost and the parent cell must guess, replacing the damaged section with a new sequence of bases. In these ways, the offspring may inherit DNA startlingly different—crucially different—from that of its parents. Such a mutant may be a "sport," such as a freak with three legs. Or it just might incorporate a new pigment into its retinas, enabling it to spot insects that are twice as far away.

Some species are very "successful," being represented by many individuals spread over a large area. In the J-curve for their community, these species occupy a position far out on the abundance axis. Mutant forms in such a population tend quickly to drown in a vast gene pool, unless they have arrived at an innovation so amazingly effective that the descendants of the mutant overwhelm the rest of the population, outcompeting it. New species may arise in this manner from time to time.

But far more commonly, it is the relatively unsuccessful species that give rise to new organisms. A species may have a small population confined to a limited area (even though it might occur elsewhere in great numbers). In the J-curve for that area, the species hovers on the edge of extirpation. Yet when a new mutation arises, it spreads far more readily in the smaller population. The innovation may enhance the survival of the species, or it may work slightly against it. Who needs a warty skin? Then comes the big change.

The weather shifts from warm to cold, a long-term change that persists for centuries. Or a previously dry area becomes wet or flooded. Or a savanna becomes forest. Or a lowland becomes an upland. Or an animal previously unknown in an area

invades it, becoming abundant enough to alter the environment for other species in the process. The possible changes are
myriad.

The fossil record reveals that most of the great radiations of
new species occurred during times of profound environmental
change. Dinosaurs, those superbly adapted creatures, all died
in the great winter wrought by the asteroid collision that
marked the end of the Cretaceous period. Mammals, with their
furry covering and superior systems for regulating their body
temperature, survived the cold. They were pre-adapted to a
chance event without, of course, intending to be. They were
lucky the asteroid arrived. No asteroid collision, no mammalian radiation.

Every epoch, major and minor, is marked by the complete
disappearance from the fossil record of whole species—and
sometimes of genera and even families. In the subsequent
epoch new species appear, one by one, often with no obvious
precedents, as though the populations that gave rise to the
new forms were simply too small to show up in the fossil
record.

Although we like to think that evolution has produced a
succession of more "progressive" forms, this is not always the
case. Before the last ice age, when Hungry Hollow was a lush
plateau filled with succulent vegetation, the saber-toothed cat
Smilodon stalked its forests. It hunted the giant herbivore
Megatherium. When the weather changed, and with it the vegetation, *Megatherium* became extinct. So did *Smilodon*. But
could we herd a pride of lions into a time machine and transport them back to that warmish clime and time to watch our
modern, "highly evolved" cats outperform the *Smilodon?* Even
if we could, they couldn't. They would have no way to bring

down such immense game. They lack the immense stabbing fangs needed to bleed their prey to death. The lions might be extinct within a year.

If the low end of the J-curve marks the grave of evolution, it also represents the cradle. Species do not evolve to meet environmental challenges. There is rarely sufficient time. They must be pre-evolved to some extent; otherwise, they become extinct. They must, in a word, be lucky.

It is the genotype, embedded in the immense word of DNA, that gives the phenotype stability. Deeply conservative, it yields up change gradually, only when bullied by mutation and crossover. With good reason it resists change, for most change is fatal. Yet not to change may be fatal too, and often it has been. If the genotype inhabits a species at the high-abundance end of the J-curve, it may be unable to change, trapped by its very success. It becomes a matter of some luck that a species should find itself at the low-abundance end (along with many others). And it becomes a matter of greater luck that one such species should be altered by a favorable mutation, whether it becomes favorable in the present or in the future.

The Book of Kaolinite

Beneath the meadow plants, beneath the forest trees, beneath the creek, lies clay. Thick, greasy, obdurate clay. Dug from the shallows of Hungry Creek, it feels slick to the touch. When our fingers slide across its moistened surface, we imagine it to be a gel, a goo—unstructured like mud. But the slickness is an illusion caused by our immensity. The merest touch sets thousands of tiny crystal sheets sliding one upon another.

In World Eight a crystalline jungle of geometric forms lies buried in the shallows. Here are families, genera, and species of clay. Their forms vary: plates and coils, diaphanous sheets, filaments and tubes. Some of the plates are glued helter-skelter into card houses. Others form books that look as though they're growing. Their pages, printed in atomic letters, have an interesting passage or two. Who could possibly guess the subject? Is it the origin of life?

The book belongs to a very common family of clays called kaolinite, the stuff that is fired to fill china cabinets. The pages

"Here are families, genera, and species of clay."

of the book are fine, crystal sheets barely a hundredth of a micron in thickness. Each page has five distinct layers, and each layer is composed of atoms of a specific type, interlinked by bonds (strong or weak) within and between the layers. From top to bottom, the layers consist of oxygen, aluminum, oxygen, silicon, and oxygen once more. Within each layer, the atoms have arranged themselves into open hexagonal arrays through which lower layers are clearly visible. Viewed through a page, the atoms of the five layers make a pattern of hexagons in which hydrogens also abound. The hexagons intersect, overlap, and interlock. The sheet is a crystal, a (nearly) perfect crystal.

The page is not rectangular like a book page, but polygonal in shape. The edges follow the hexagonal lines dictated by their crystalline geometry. Every page of this strange book has exactly the same shape bounded by the same purposeful polygon. The book has no spine, but the pages bend gradually, as though their number were continually growing. And in fact, they are. The book of kaolinite expands slowly under the currents of Hungry Creek. Some would say it is alive.

The uppermost page in the book of kaolinite is currently recruiting new units. Metal ions and molecules of silicic acid, the basic building blocks of all clay minerals, drift past the page. Most of the units have come from less perfect neighboring species in the silicon jungle, now dissolving. Some units have dissolved from clays upstream, and a few have leaked from the ultimate source of all clays, weathering granite. Glacial boulders up the valley make modest contributions with every rain.

The silicic acid units consist of a single atom of silicon linked to four oxygens, each of which bears a single hydrogen atom. The molecule has a tetrahedral form, forerunner of the hexagonal structures. The metal units are ions, molecules with an attitude. The single atom of metal, whether aluminum, iron, or potassium, lacks electrons. The resulting positive charge attracts the negative side of surrounding water molecules, forming hydrogen bonds with them. Thus hydrated, the metal clings to as many water molecules as its charge will hold.

When the silicic acid units and hydrated metal units have the right abundance, they may not only encounter a growing page in the book of kaolinite but become part of it as well. The page may be incomplete, a thin crystal plateau. One by one, the units slip into place on the edge of the plateau where there are spots for atoms of oxygen, silicon, and aluminum. They repeat the pattern of the previous page—not only the simple, unvarying hexagonal geometry but also a subtler pattern called a defect.

When a unit joins a site on a crystal page from the surrounding solution, it must choose among three positions within the site. The positions occupy the corners of an equilateral triangle. Which position a unit chooses is influenced by neighboring units that have already joined the crystal, whether below the layer currently a-building, or within it. The new units therefore arrange themselves into domains in which all the new atoms occupy corresponding positions within their sites. To the eye, they constitute a unit, an aesthetic whole. Yet there are other domains in which the new atoms have all made another choice. The pattern of domains determines much of the crystal as a whole, including the shape of the page.

At some sites within a page of kaolinite, certain metals may be preferred. Here an iron might be more readily accommodated than an aluminum, there a phosphorus.

The pattern of domains, the presence of metallic atoms, and other irregularities are transmitted from layer to layer, sheet after sheet, by small differences in the force fields that surround each site, like the pea the princess felt through several mattresses. In these irregularities, faithfully copied by crystal growth, the clay has something resembling a genetic code.

In some clays, the pattern of defects and other repeated irregularities may determine the fate of the clay. In the book of kaolinite they may control the size and shape of the pages, how soon the crystal will cleave, and how well it will resist dissolution. These and a myriad of other factors can be imagined, each adding to or subtracting from the kaolinite's survival and reproduction.

As long as the water that permeates the bed of clay retains the same composition, new units will be added to the book. Presently, the stack of new pages will obtrude into the mildest of currents that move between the grains of sand above the clay. A few pages will break away to drift slowly between the grains, perhaps out into the great current of Hungry Creek itself. Downstream, the nanospore may begin a new book somewhere on the bottom. Its pages will have the same shape and size as those of the parent "organism" because the pages have the same pattern written on them. Each page is a naked "gene." Phenotype and genotype are one.

When saturation levels alter with the flow of the creek, everything changes for the buried clays. Species that had been growing may begin to dissolve, giving up their units to the currents once more. Species that had been dissolving may begin to grow again. Clay has always done this. Does its lifelike behavior hint at some ancient secret?

The Book of Kaolinite—Chapter One

On the primordial Earth four billion years ago, proto-continents had already formed from hot, magmatic up-wellings. The new rock was granitic, studded with crystals of quartz, cemented by grey feldspar. All met a dark sky, heavy with water vapor, nitrogen, and carbon dioxide. There was no free oxygen to breathe and no one to breathe it. One of the most common elements on Earth, oxygen, was all locked up in the rock and in carbon dioxide.

Rains fell on the newly emerged rock. They fell for forty million days and more. Their acids dissolved the feldspar readily. The quartz crystals washed away as sand.

Hungry Hollow began as a hot, barren moonscape drenched by a continual rain. There, and in all the surrounding countryside, the rains formed uneasy lakes or rushed in wild rivers to the sea. In this vast laboratory, clays formed everywhere and in every conceivable environment. They lined stream beds, made vast layers in lakes and sea bottoms, and seeped through pores of sand in newly forming soils.

By the end of the great rains, every species of clay we know today had formed in one place or another. There were kaolinites, smectites, montmorillonites, illites, Gibbsites, chlorites—a multitude of families. Within these families were definite and indefinite species that appeared, disappeared, and then reappeared elsewhere. All the clays that formed were, by the laws of statistics, inevitable, but some were less inevitable than others.

The simplest clays formed first. They fed on the everplentiful stock of silicate units that washed daily from the igneous rock. They were hardly species, making hit-and-miss, uncoordinated structures in which crystals rarely replicated from one layer to the next. And when the saturation level dropped slightly, they began to dissolve, feeding in turn a new generation of more precisely formed clays in which the crystalline sheets repeated with greater regularity. A kind of evolution had already begun.

Among the newer clays were those for which "survival" could be a meaningful term. Their crystalline patterns replicated faithfully from one sheet to the next, and when they became large enough to break apart, the freshly exposed surfaces grew anew. Many of the new species of clay had a special affinity for organic molecules. These clays grabbed carbon compounds with ease, adding them around their edges to protect them from dissolution, or incorporating them into grooves and channels within the more exotic geometries.

As Earth cooled and oceans formed, the heavy clouds gradually dispersed and the Age of Sunlight began. Some of the organometallic compounds thus incorporated became catchers of light, early photosynthetic molecules that transmuted photons to spot-weld the unwieldy carbon compounds into more complex units. The operation split a molecule or two of water, creating a pair of oxygen atoms: the O_2 molecule. Gradually, the atmosphere filled with oxygen, even as levels of carbon dioxide dropped.

At first, the primitive carbon compounds did only physical work. Their powerful bonds meant that they did not break easily. As chains, they served as holdfasts to prevent washout; adhering to exposed faces in adverse conditions, they prevented dissolution or enhanced the crystallization process. Being strong, the bonds of organic molecules also served as energy stores. Formed at sites of photosynthesis, they were channeled by the glassware of exotic clays to controlled reaction sites where their bond energy could be released for the synthesis of other organic molecules. They became enzymes too, speeding up the synthesis. They became proteins. They became, ultimately, ribonucleic acid: RNA.

In the end, the clays made a fateful decision, one to which they were inclined by convenience alone. Increasingly, the synthesis of new compounds was directed by RNA. Gradually, the patterns stored in the crystal replicands mattered less and less, as more and more information was stored in the new, high-tech proteins. There was a genetic takeover. Lipid membranes replaced silicate ones. Deoxyribonucleic acid, DNA, emerged, along with the first self-reproducing cells, the archaebacteria.

In the end, the cells became independent of the parent clays. They cornered the market in organic molecules, and the photosynthetic clays disappeared. Only the pre-photosynthetic clays remain, ready to begin anew should the organic world falter and die.

Didelphis
virginianus

She is called by many names but answers to none of them. In various languages she has been called "incredible mother," "two-bellied beast," "fox-monkey," "bush rat," "fetid one," and "white-face." The Delaware language claims the honor of providing her common name: "a'passum," a guttural preceding the word *possum*. It means "white animal." Greek, with its unavoidable scientific ring, declares that she has *di delphis,* or "two uteri," there being two wombs that supply the birth canal. (Her mate has a cleft penis.) All these names collectively convey her qualities, but she is Virginia to her friends.

Like Lotor, Virginia is an ancient being. But she is twice as ancient, her likeness having been present at the founding of the mammals in the Jurassic period. To look at Virginia's face is to peek through a window in time. We did not know the dinosaurs, but we know someone who did. A possum-like creature peeked out at *Ankylosaur* from behind bushes when the *Didelphid* line began during the Cretaceous period in South

America. She scavenged the remains of small dinosaurs and was eaten, in turn, by large ones. She was there when the great asteroid struck Earth, bringing the dark, astronomical winter that killed the dinosaurs. She survived to witness the evolution of her sister mammals into many forms with hoof or claw. When South America met North America, she migrated north, meeting Lotor's earliest look-alike, the dog-coon-bear of the North American Miocene epoch, some thirty million years after the birth of the mammals.

Lotor is not yet aware that his great aunt prowls the vines of wild grape that mound the bushes where forest meets farmland south of Hungry Hollow. A mild and occasional shake of the mound betrays her presence. She clings to a vine with all four feet and with her tail, maneuvring herself close to snatch the small, semi-sweet grapes in her mouth. On her back cling six more possums the size and shape of small rats.

Two months ago she gave birth to nine young. They had spent a mere thirteen days in the womb, emerging as precocious worm-like fetuses that made an epic crawl from the vaginal opening, up the fur of her belly, to a pouch where thirteen nipples awaited them. Thirteen days and thirteen nipples brought bad luck to two of the fetuses. Seven made it to the doorway in her belly, crawled inside the pouch, and found nipples that they took into their mouths, remaining attached for weeks. But two became lost in the confusion and did not make it to the pouch. Some days ago, a third grew weak and, giving up its teat, slipped unnoticed from the pouch while Virginia crossed Hungry Creek.

The pouch, or marsupium, is the next clue to Virginia's ancient lineage. She is the only marsupial to inhabit North America. Her marsupial ancestors of the Jurassic period inhabited

Gondwana, a vast southern continent that slowly broke up during the Mesozoic. The first split saw a future Africa and India drift away, leaving three more continents still joined: Australia, Antarctica, and South America. Some of the marsupials lived in Australia, some in the future Antarctica, and some in South America. The second split, following hard on the heels of the first, took Antarctica toward the south pole, south America to the west, and Australia to the east. By the time Antarctica had settled over the south pole, its pouched mammals had been extinguished by cold. Yet marsupials remained in Australia and South America. Even as Lotor's distant forebears pursued their general way of life, South America finally joined North America. Virginia's direct ancestors crossed the newly formed land bridge. They have since disappeared from South America, leaving it to many sister species.

In the survival of Virginia herself are clues to the survival of her kind. First, she is Jill of all trades, mistress of none. She will eat anything of nutritional value, including fruits, seeds, nuts, insects, carrion, and even the dung of other animals. She out-generalizes Lotor himself. She climbs as well, runs as fast, lies as low, and prowls by night or day.

Second, she is not proud and does not stand on her pedigree. If threatened by another animal, she may hiss horribly and show a very long set of fifty teeth, or she may run away, or she may humbly submit, slipping into a cataleptic state. Her eyes glaze over, her tongue lolls, and her body becomes stiff and lifeless. She lies like one dead. She smells like one dead. The deception works on all but her carrion-eating colleagues. When the predator wanders off in search of livelier prey, Virginia slowly rises, looks about, and then tries to remember what she was doing before she was so rudely disturbed.

Playing possum, she seems to say, "Go ahead. Eat me. There are plenty more where I came from." And so there are. With up to three broods a year, the possum can outproduce any other North American mammal of her size. Virginia is on a roll these days. Her family has recently invaded Canada.

As she munches on the wild grapes, Virginia reveals a set of unspecialized teeth that echo a Jurassic reptile, the therapsid that was ancestral to all mammals. Similar jaws are found in the insectivores, a large order of placental mammals that are nearly as ancient as she. Her tail is bare and scaly; her ears and paws are also bare. Her skin is pink, the unpigmented color shared by many mammals, including a race of humans. Although her extremities are naked, Virginia dresses the rest of herself in thick, cottony underwear, grey with hints of yellow and buff. Over this dainty garment she wears a thin coat of white guard hairs.

Her feet are almost hands. She has made tracks in the soft mud of a point bar on Hungry Creek. They reveal forepaws with five forward-pointing toes well spread out and larger than Lotor's. The prints of her hind paws are strange indeed. Four digits point forward, but a fifth points inward and to the rear. The rear thumbs are opposable, enabling her to grip branches like a primate.

She may hang by a back leg or two while feeding. She may even hang by her tail, but not for long. This is sport for the young, whose tails more easily support their weight. As though to prove the point, one of her babies has left its spot on her back. It hangs upside down from a horizontal length of vine to watch its mother feed. It looks through small black eyes set in the white face it shares with her. It is a true babe in the woods.

Virginia looks slowly around for more grapes and then suddenly pricks up her ears. She makes a barely audible clicking noise that brings the stray child to attention. It obeys instantly, scrambling once more for the security of its mother's back. Virginia makes her way ponderously down the bushy arbor with much swinging and swaying, her urchins lurching this way and that. They clutch tightly to her fur with tiny paws and tails. They look about in wonderment at the passing leaves. Their sharp ears now pick up the distant panting of a running dog, one following a trail of scent.

On the ground, Virginia heads for the nearest tree just as the dog, a German shepherd, bursts into view. She cannot play possum this time, for the dog will eat her children. When she reaches the tree, she begins her climb, scuttling up the rough bark as fast as she can. The shepherd growls fiercely and makes a dash for her, leaping at the last minute to seize her in its jaws. But she is already out of range and the dog rages about the base of the tree. When she reaches the first branch, she rests and looks down, knowing she is safe. Her children also stare down at the dog. They will learn from this lesson that dogs are to be avoided, but they will not always do what their mother does next. She defecates a scented rope that hangs in sections and then falls to the ground, narrowly missing the dog. It is her comment on the "higher" mammals.

Lotor has heard the commotion, but the noises have come from the forest side of the hackberry. Carefully, he leans out, grips some bark, and edges around the trunk far enough to see the dog. He knows the German shepherd. It will whine and bark for a few minutes and then wander off. A well-fed dog is never a serious hunter. But Lotor also knows the dog makes these sounds when he has treed a raccoon. He concludes that another raccoon has entered his valley. He feels a raccoon's anger. He will investigate later, when the sun goes down.

Lotor's brain has four times the volume of Virginia's. In his position, she might not draw such extensive conclusions. She might not deduce that another possum has entered the hollow. But she doesn't need such deductions. For one thing, she does not particularly care whether there is another possum. Every animal knows only what it needs to know. Virginia lives only for the moment, and for the moment, she lives in Hungry Hollow.

Animal Minds

With the possible exception of Dianne, Lotor is the chief intellect of Hungry Hollow. A behaviorist would make him a stimulus–response engine. A cartoonist would make him human, endowing him with word balloons. Both schools of thought shortchange the real animal. Somewhere between lives the real Lotor, moving and being.

The interpretation of behavior is not a science but a perilous art. Nothing can be proved. Yet charges that Lotor cannot think, feel, or imagine can be tried, as in a court of law. And if some evidence is lacking, careful observers (neither behaviorists nor cartoonists) can be called on to render an opinion. Consider what inward events might produce outward incidents before assuming that animals have no minds of their own.

Memory

As he wanders the Hollow, Lotor keeps several apartments and refuges in mind. Although he prefers to sleep in the hackberry these days, he also keeps an apartment in a large silver maple at the edge of the meadow north of the hackberry. The protection from the elements is not complete there. The crotch of the tree has rotted out, and there is a roomy space into which Lotor can disappear, leaving neither sight nor scent for marauding hounds. But the apartment lets in rain and is suitable only as a fair-weather bivouac. Lotor does not stumble upon this apartment, rediscovering it with each visit. He remembers the tree, knows where it is, and can go there directly from anywhere in the Hollow. And when he arrives at the tree, he recalls the best route up its friable and dangerous bark. He has in his mind a map library of sorts.

More than this, much more, Lotor knows and remembers the best places to look for earthworms, crayfish, wild plums, field strawberries, turtles and turtle eggs, bird eggs, arrowhead roots, beetle larvae, madtoms, clams, snails, baby birds, hickory nuts, wild grapes, snakes, frogs, toads, salamanders, land snails, grasshoppers, cranberries, puffballs, water, beech nuts, honey, crab apples, baby snappers, maple sugar, fiddleheads, tree frogs, water lilies, hawk spoils, and cicadas.

Memory is knowledge. Some memories involve particular knowledge, such as the delicious spotted salamander he ate the other day by the old log. Other knowledge is general. "There's no point in checking the old log for salamanders on all rounds because salamanders are not often found there—or anywhere." Other general knowledge involves the fortuitous nature of general foraging. You never know what is going to turn up when you prowl the shallows or along the hawthorns at the forest edge! But something always does. That is the general knowledge that Lotor needs. The rest is useless. He does not need to know, for example, that humans never have two heads. Should a two-headed Dianne appear one evening by the creek, he would not find it remarkable.

Planning

Lotor has an itinerary that he follows night after night. It is half about food, half about information. He is drawn to the places where food is. He visits the nearest place (unless a farther one draws him more powerfully), then the next nearest. His plan is not detailed, but it is firm enough to prevent him from zig-zagging all over Hungry Hollow, consuming valuable energy instead of food. He also approaches every place with the nearest refuge in mind, from the old hackberry to the hollowed-out log by the creek where he recently found the salamander.

And should he have his fill of nuts or crayfish, he will still visit the other places just to determine the state of things. Near the upper creek there is an old willow with a hole beneath its roots. He checks it regularly, as a matter of course, to see whether the striped skunk still lives there. He will not add it to his current list of refuges until *Mephitis mephitis* has left. Lotor feeds his body with food, his mind with information. To stay alive, he must keep his itinerary up to date.

Imagination

To keep track of his apartments and to run unerringly toward one that he cannot see yet knows is there, Lotor requires nothing less than a mental map of Hungry Hollow—a way to visualize the place and to pinpoint himself on the map: "I am here; the hackberry is there." The map, whatever its form and however rudimentary it is, points to an imaginative facility possessed not only by Lotor but to some degree by all animals. If the strict behaviorist requires that no thought, let alone imagination, come between stimulus and response, then Lotor's brain must be the size of Hungry Hollow itself to store all the tables that will tell it what to do in the infinity of situations that *might* arise in the course of a single day.

What is imagination but the ability to simulate reality? A brain that can simulate, building even the most rudimentary model of reality, has enormous advantages over one that cannot. A simulating brain can foresee consequences, can plan, can imagine. It will approximate reality by mental tokens that behave enough like their real-world counterparts that reliable conclusions can be drawn. The behaviorist may not realize that simple imagination comes relatively cheap, inhabiting Lotor's small, smooth lobes with room to spare. Does Lotor not imagine himself fishing in the shallows before he visits them? Does he not form some mental image of the place, though that image be crude, like the sketch made by a child?

As for thought itself, there is no doubt that Lotor makes deductions. Witness his recent (faulty) conclusion that another raccoon had invaded the Hollow. He could easily have been right. He certainly remembers such an encounter, though its memory has faded to a shadowy précis.

Emotion

Lotor remembers, albeit vaguely, looking up from business under the flowering dogwoods in the early dawn of a spring morning to see another raccoon nosing about his beloved shallows. Anger gripped him. What else called forth the low, barely audible growl that rose in his throat, mixed with a whine of frustration?

If the other raccoon had seen Lotor first, would his reaction have been the same? Not at all. He too could think, and he would have known that this was not his territory. The intruder felt fear and stopped what he was doing to watch the oncoming Lotor. It is a strange habit of raccoons to appear stupid in the middle of an emergency, standing still to think the situation through. As Lotor rushed up, the raccoon in the shallows still had time to turn tail, dash across the creek, and turn again to see whether this barrier was enough to stop the furious one. (Other raccoons, faced with oncoming headlights, often give themselves too much time to think about a subject they do not fully understand.)

Lotor screamed as he came to the shallows. He screamed because he was angry, very angry. Lotor's rage had done its job. His adrenaline was way up, and his heart beat like the drum of a flicker. His mind ran at a new pitch on a flood of glucose hastily dumped into its arteries. On the warpath, his brain reorganized itself around the logic of fight-or-flight. Lotor was not then conscious of the logic. When Lotor howled in rage, he did not say to himself, "I am howling in order to give the intruder a chance to retreat and therefore spare us both the trouble of a pitched battle that neither of us wants."

Tearing across the meadow, he caught a glimpse of the other. It looked up and paused, as though uncertain what to do. Lotor saw the uncertainty (why should he not?) and felt a burst of confidence. But when the intruder stopped on the other side of the creek, Lotor was surprised anew, then frightened, then newly angry. Upping the ante, he prepared to fight to the death. It would be holy war. Food and territory are sacred.

He growled again and then lowered his head until his chin almost touched the ground. He depended implicitly on the other recognizing the preparation for a raccoon rush—that terrifying charge that so belies the peaceful waddle and comical gait. In the rush he would be grey lightning, committed to taking hold and never letting go until someone died. In the rush, the other would have less than a second to decide the great issue for himself: fight or flee. The intruding raccoon turned tail just as Lotor was about to pull the trigger. Relieved, he watched the retreating figure, waiting until it had climbed nearly to the top of the ridge.

Then he growled softly. It was dawn. Lotor returned to the silver maple, where he had been staying those days, and slept the sleep of the righteous. For half an hour his sides moved gently in and out. Then his legs started to churn and he whimpered in fear. Had the intruder reappeared, larger than life? Did it chase him even then, howling with its own rage. And did Lotor's refuges disappear, one by one, before he got to them? The simulator must practice long and often.

Permanent Clearcut

It is the first day of autumn. Lotor has been sleeping in his hackberry apartment since sunrise. Slowly at first, but insistently, a banging noise creeps into his consciousness. It is rhythmic: Bang bang bang bang. Then it stops for a while. Then it starts again. Bang bang bang bang bang. He has heard this sound before, but never so close. He is alarmed and sits up, the better to listen: Bang bang bang. He puts his head outside and grows more alarmed. The sound is not coming from the source of all noise, the Place of the Dogs, but from the opposite direction. The sound is coming through the forest.

Something subtle shifts in his brain. A decision process has begun, one that will change the rest of his life. It has to do with his perception of "here" (Hungry Hollow) and "there" (Whispering Pines). The two have begun to meld. The banging has blurred the distinction. Although the banging stops, Lotor listens for a long time. It is late in the day. He cannot sleep.

The human colony called Whispering Pines consists of 26 box-like structures fabricated by the humans themselves with the help of machines that burn gasoline or run on electricity. Birds make nests, groundhogs make burrows, bees make hives, muskrats make lodges, but only humans make boxes to live in.

And only humans make large, complicated machines that work tirelessly for them. They have welcomed the machines into their lives, believing that the machines will liberate, rather than enslave, them. Some of them have adopted an implicit religion, a self-fulfilling prophecy that machines amount to a vast pupa from which the human larva will emerge transformed. The human butterfly will crawl forth from the infrastructure of cogs and wires to spread its prosthetic wings. Co-evolution.

The machines *have* evolved after a fashion, and now exhibit their lineage in their morphology. The nineteenth century saw the first great radiation of machines of the phylum Dynamopartida, machines with moving parts. Within this phylum the class Itineromachina, or moving machines, proved most useful to the human species, and their further evolution was encouraged. The order Carbomata, steam-driven machines, evolved into many different families, genera, and species, a few of which are still with us. Around the turn of the century, the second great order of Dynamopartida evolved. The Oleumata, or oil-driven machines, displaced the earlier Carbomata, displaying at the same time a remarkable parallel evolution in which, for example, the oil-burning train replaced the coal-burning train.

Whispering Pines would have been impossible without the cooperation of such creatures. In the beginning—not so long ago, but before Lotor's time—ten hectares of trees were felled

"... ten hectares of trees were felled by chain saws, small machines of the family Oleuminidae."

by chain saws, small machines of the family Oleuminidae. When the ground had been cleared, large, lumbering beasts of the family Catenidae appeared. The family name reflects the circulating tank tread on which the machine moves. The bulldozer, *Batillum status,* leveled the earth, snorting and clanking over all obstacles.

The power shovel, *Batillum mobile,* squeaked and rumbled to the site of each house in turn. A large scoop in front of the machine bit into the earth. The machine snorted with sustained rage, drawing the shovel with powerful hydraulic muscles, clawing at the earth. With more creaking and rumbling, the machine swiveled on its base, the bucket swinging grandly in World Zero, lurching to a halt in mid-air. Another machine called a dump truck, *Vehiculum cista,* its transmission whining submissively, backed gently until the box on its back was under the shovel. The shovel tilted and the earth poured into the box. The dump truck settled a little, uncomfortable.

When the foundations had all been dug, the power shovel began another series of holes for septic tanks.

Humans scrambled like slaves inside the new foundation, putting up slabs of laminated wood to contain the next stage of construction. Then a new machine arrived on the scene. *Vehiculum dolium,* the cement truck, belongs not only to the same family (Hyporotidae) as the dump truck, but to the same genus as well. It is also a truck, but instead of a box on its back, it has a large, rotating cylinder filled with stones, sand, lime, and water. The barrel turned and turned, the mixture churning within. The cement truck backed up to the foundation amid shouts from the humans. They pulled a long trough from the body of the truck, swinging it out over the expectant plywood. At the flip of a handle, wet cement streamed out of the barrel and down the trough, settling into the forms—a rich, grey elimination of half-digested gravel.

As the foundations hardened, another machine dug a network of trenches. The backhoe, *Carpentum trulla,* belongs to the same family as the trucks. Yet the action of its small, rear-mounted bucket closely resembles that of the power shovel. Into the trenches, humans laid water and gas lines.

Lotor's great-great-grandmother saw the whole thing. She watched the humans put up wooden forms, run wire, and install plumbing, flooring, insulation, drywall, siding, and roofing. She heard the whine of their power tools (the family Electrominidae). Then all was quiet. The forest behind the ridge had nearly disappeared, replaced by 26 large boxes. Permanent clearcut. Well, perhaps not quite. The developer's love for nature finally found expression in the pines for which the colony was named. A few dozen Eastern white pines were planted in the colony, one in front of each house and a few more along the road. The developer's love of nature was not matched by his knowledge, unfortunately. Pines prefer sandy soil and do not do well in heavy moraine clay. Often they weaken and fall prey to various diseases. The whispering white pines were doomed. Later they would turn a ghastly reddish-brown, attacked by the white pine blister rust, *Cronartium ribicola.* When they whispered no more, the clearcut would be complete again.

The colony quickly filled with humans who arrived in cars, *Vehiculum vulgaris.* They brought their children. They brought the dogs that, according to Lotor's great-great-grandmother, ran things in Whispering Pines, endlessly barking orders.

Now it is Lotor's turn to bear witness. As evening falls, he climbs the ridge, hidden by bushes, until he reaches the top. He moves cautiously through the thin line of trees only to find chain-link and board fences. He prefers the board fences, for he may climb unseen. When he gets occasional whiffs of dog, he avoids traveling upwind. He finds a garbage pail with an intriguing aroma that reminds him of the sandwich. He climbs the pail to investigate the top, inserting his clever hands under the rim, prying it up. There is resistance. He pries harder. The lid springs off, surprising him, and the pail lurches and falls, hitting a fence. Voices sound inside the house and a back light clicks on. "Who's that?" Lotor is over the fence in a trice. There is much to learn.

He has already learned that garbage pails are the best source, but garbage bags are better. These can be readily disemboweled, spilling many juicy scraps of bread, discarded bones and fat, potatoes, and dish scrapings. There is a well-balanced diet here. He learns something else on this night, too. Be careful when you fish in tins with attached lids still inside them. If you reach for morsels past the lid, it swings against the hand that seizes, seizing it in turn. Lotor panicked and pulled his hand, screaming at the pain of lost skin. Dogs set to barking all around him, and he ran for the Hollow, scrambling over fences, ignoring the pain in his paw. He ran down the cliff, across Hungry Creek, and back to the forest.

Only at home does he attend to his paw, licking it over and over. The night is still young, but he will not go out again. As the throbbing in his paw eases, he buries his nose in his tail and drifts into a fitful sleep. He has a strange, disquieting dream.

First come yellow monsters that roar. They have black stones under them instead of legs, and they crawl by rolling their stones. One monster has a large claw that digs a hole in the earth. Then comes a new monster with a big yellow stone

on its back that keeps rolling without going anywhere. Groaning, it evacuates into all the holes. Then come the two-legged
animals, crawling over everything and screaming at each
other. They put dead trees around the holes and attach more
dead trees. Then they put colored bark over all the dead trees
but leave some holes in the bark, which they cover with a kind
of ice. They look out from behind the ice, watching dogs that
come and go.

Ursa Major

It is a cool morning in early fall. A light mist covers the floor of the forest. At the forest edge, beside the county road, someone has recently nailed a sign to a tree:

HUNGRY HOLLOW
WHERE PEOPLE
AND
NATURE MEET

Beneath the sign, a chain saw rests against the tree. It is an elegant machine, shaped by inorganic evolution (in which form follows function) into an organism perfectly adapted to its environment. It is a thing of the earth that has erupted to the surface. Its metal was smelted from buried ores of aluminum and iron. Its gasoline was refined from oil of fossil plants.

A cast aluminum two-cycle engine and gas tank together occupy a yellow housing. A tubular handle encircles the housing above, and a second, pistol-grip handle is mounted directly on the frame. A third, smaller handle dangles from a pull cord that runs inside a circular hump on the side of the housing. This cord winds around a pulley that shares the engine's crankshaft. A long, shiny blade with a groove on its edge extends forward from the housing. The groove makes a track for an endless chain that runs around the rim of the blade and back inside the housing, where a gear engages the inside of the chain, as on a bicycle.

The outside of the chain bears teeth, one per link. The teeth on consecutive links are attached to opposite sides of the chain so that when the machine is running and the chain begins to whip around the blade track, the teeth cut into wood like so many miniature axes, stroking first on one side, then on the other, kicking the chips out of the cut.

The engine has one cylinder with a spark plug and a piston inside. When a spark leaps the gap at the base of the plug, it sets the gasoline vapor in the cylinder on fire. The fire is brief and explosive. It cannons the piston down and turns the crankshaft a half-turn. Angular momentum brings the crank around for another turn. The piston returns to the top of the cylinder while a small port opens to exhaust the burnt vapor. Another port admits fresh vapor as the piston descends the cylinder again, and another spark leaps the gap when it comes up.

Someone in a plaid shirt and a yellow hard hat picks the machine up, braces it on the ground, and pulls the cord fiercely. The machine growls like an angry animal and then roars several times in succession as someone squeezes the trigger to test the throttle. The chain is a blur. The blade rests against an 85-year-old beech, and someone squeezes the throttle again, pushing the blade gently against the smooth grey bark. A notch appears as though by magic, a notch that sinks without effort into the heart of the tree.

Dianne hears the chain saws running, throws on her coat, and dashes out the back door of her house. She almost slips on the wet cliffside path. She is out of breath by the time she has run across the meadow and through the old forest.

"What are you doing?"

One of the cutters cocks his ear but keeps the chain saw running.

"I said, What are you doing?"

The man reluctantly kills the motor and signals his partner. The forest is momentarily silent.

"What are you doing?"

"What does it look like we're doing?"

"You can't cut those trees."

By way of an answer, the man points at the sign and, without waiting for Dianne's reaction, starts his saw again.

"Jesus Christ!" She is beside herself, with barely enough presence of mind to be astonished at what she has just said. The men don't hear her. She runs through the forest again, across the meadow, and up the cliff. She stands in her kitchen, bent over, nauseated from the run, gasping for breath. How can this be possible?

It takes three telephone calls before she reaches the right department of the County Administrative Office. A voice tells her that the Hungry Hollow Condo Project was launched some time ago and that the required notices were mailed to everyone in the area last spring. "Well, I didn't get one!" She hangs up without saying goodbye, something she has never done before.

There is nothing she can do about it. The workers will fell a tree every twenty minutes, she thinks. All she can do is watch.

It comes to her, just then, that it was like this for the native people: the sound not of chain saws but of axes. It comes to her that they too were powerless. It comes to her, perhaps, from the cookie tin that sits on a kitchen shelf. Her eyes are wide for a moment, and then she rushes to the telephone book. Within ten minutes she has started her old Ford.

The sky is grey as Dianne drives north on the county road, the sound of chain saws slowly fading behind her. A few tentative drops of rain streak the windshield. The wipers squeak and bang, squeak and bang. She presses the accelerator pedal gingerly, watching the abstract needle point to higher numbers. But the car does not seem to go any faster. It is like a slow-motion nightmare.

Two hours and two wrong turns later, the car rocks and pitches in the potholes of an old road that turns abruptly to reveal a sign.

A'NOWAGHI FIRST NATION

Below the lettering, someone has spray-painted a crude day-glow message: *Watch Your Ass.*

Dianne wonders whether this is a serious warning or just the work of a prankster. She drives very cautiously past some poor-looking houses with dirt lawns, a tricycle lying on its side, a black and white dog chained to a porch and barking furiously. A small girl stands by the road, sucking her thumb and staring with great seriousness at the passing Dianne.

The band office is not a reassuring structure. There is no power here nor any pride, only weariness. The duly elected chief is in, but he's arguing with a woman in his office. He becomes aware of Dianne after a few moments. "What?"

"I need help. I have something here, but I need your help. They're cutting down the forest, you see."

"Hold on. Hold on," says the chief. He is large and fat. He pulls his bulk out of the armchair and approaches the desk where Dianne has put the cookie tin. "What're you talking about?" The woman also approaches the desk, her argument forgotten.

Dianne opens the lid, carefully removes the bag, and pulls out the long, thin bone. The chief laughs uproariously. "What do I want with a bear-prick?" Dianne is confused, lost in a world that speaks a different language. The woman takes Dianne's arm suddenly and decisively, snapping at the chief. "Albert, this is a woman thing! Let me handle it."

Dianne sits at an oilcloth table. The window to her left frames two other houses, one painted turquoise, the other an odd shade of brown. A line of wash hangs soaking in the rain. Across the table, the woman called Mary Bear examines the items in the bag.

"It's a good thing I got you out of there in time. That was Albert, head of the Band Council. I don't think he has a clue what this actually is."

"What is it?"

"Well, it looks like a very old bag. Let's see what's in it. . . . Here's what Bert was laughing at. It's the penis-bone of a bear. To some people it's funny, but not when it's in an old bag like this. No way. What else we got?"

Mary withdraws the colored skin. She whistles. "A hummingbird! Mercy, mercy. That's no bird from these parts. This has got a very scary feeling about it. It has power." Next she

214

removes the seashell. "And this here's a Megus Shell, traded all the way up from Florida. And this stone with a ring around it. That's a talisman of one kind or another. Yes, yes, yes." Then she pours the ten bear claws from the table.

Dianne is startled to see tears in the eyes of this gruff woman. "I swear I dreamed this was going to happen. Where in *hell* did you get this bag?"

It is Dianne's chance to talk about the cutting. She describes the forest and the grave. She talks about how old the bones seem to be. She talks about men in hard hats stomping on the grave.

"Don't push it."

Mary gets up abruptly and walks to the back door to call someone. Within the hour there are nine more people in Mary's kitchen. Some sit in chairs, others lean against the wall. The faces, Dianne thinks, are real. The ancestry is diluted but unmistakable. They are the first people still.

"Let me explain something to you," says Mary, half to Dianne, half to the others. "These days the rez is split down the middle. There's the band council taking tons of money from the government, sharing the loot with friends and relatives. Then there's the People of the Longhouse, who're trying to learn the old ways and keep them alive. That's us. We don't have any control, but our time is coming. Maybe it's here now."

She holds up the bag. "D'you know what this is?" She looks from face to face in a dead silence. Dianne senses a power in Mary's manner and marvels that the others, mostly men, hang on every word. "It's a medicine bag sent to us from long ago. It's a miracle is what it is. It's a message. It says, 'The holy ground is being disturbed.' It says, 'Those people—you know the ones—are at it again.' It says, 'The disturbance must end

"The Great Bear, Ursa Major, circles the pole, calling forth Procyon."

today.' And it says, 'This medicine will help.' That is why the message took this form. Okay?"

They all say, "Ho."

There will be a blockade of the work at Hungry Hollow next morning. The Warrior Society will be mobilized overnight. Mary then phones "the professor," a man she says can help them. She says to Dianne, "You better stay for supper."

The professor turns out to be an archeologist who arrives at the reserve just before sunset. He is a friendly man, but a little diffident with Mary, as though remembering past disputes. His eyes bulge involuntarily when Mary shows him the contents of the bag. "You can have this when we're done with it. But here's what you have to do." The professor nods at everything Mary says, including the blockade. He can't take his eyes off the bag.

"We've got him on our side, not to say hooked," says Mary with a laugh. The professor has just left. He will arrive at Dianne's house in Whispering Pines at first light tomorrow. She will show him the grave, and he will set things in motion. He will sign the form that invokes Section 45 of the Native Heritage Protection Act.

"That grave better be real," shouts Mary over the starter of Dianne's car. "Yes," agrees Dianne. "It better be."

At home again, she stands in the kitchen, feeling drained. She takes the flashlight from the cupboard, thinking she will cross the Hollow at night to check the damage. She can count the stumps. But she gets only as far as the back steps before exhaustion hits her. She thinks she would rather not see the damage, after all. So she sits, staring up at the unwinking stars of this cool night. The Summer Triangle has begun to slide from view. The Great Bear, Ursa Major, circles the pole, calling forth Procyon.

Ahead of Dianne there is a scrabbling sound on the fence. A fur-covered face appears over the top, a black mask with two hard, black eyes. Two soft, black hands with hard nails grip the boards. It is an ancient lineage, too busy surviving to be proud. *Procyon lotor* has come a-begging.

Notes

Procyon lotor

The order Carnivora evolved in two main lineages: cat-like carnivores in Asia and Africa, dog-like ones in North America. The phrase *dog-like* does not describe the North American ancestor very well. It was smallish, arboreal, and not a great runner. It gave rise to the North American families represented by the dog, the bear, the wolverine, and the raccoon. Only in the raccoon do we see suggestions of the other three families.

Raccoons that have lived solely in natural environments do not readily enter cities or suburbs. Urban raccoons use the same skills in a different setting, passing all along as "culture" to progeny. Transplanting an urban raccoon to the country is a sentence of death for three reasons: It will (1) almost certainly arrive in another raccoon's territory, (2) not know that territory, and (3) have no rural skills.

The Tippecanoe Sea

Explaining how various sedimentary formations got that way is not always easy. The Hungry Hollow formation of southern Ontario and the Ledyard Shale of eastern New York state, parts of the Hamilton Group of the Middle Devonian, have a disorganized look, with all fossil fauna askew. But this helter-skelter appearance could easily have been caused by the churning of marine worms. Increasingly, however, geologists have recognized how well the occasional catastrophe explains things.

Dianne sapiens

The greatest environmental threat today is loss of habitat for species of almost every type. Human communities, factories, and farms amount to a "permanent clearcut," in the words of environmentalist Patrick Moore, a founder of Greenpeace. While so many species become increasingly rare and then wink out like lights, mainstream bioscience pours money into molecular biology and genetics. And governments trim the scientific civil service, firing the taxonomists who are essential to knowing what species are still with us—and which are not.

The Hackberry

Some biologists currently recognize six kingdoms, not five. To the additional kingdom belong archaebacteria with their primordial biochemistry. For the sake of simplicity, we have chosen to include them in Kingdom Monera (all the bacteria), as did Lynn Margulis abd K. V. Schwartz in their book *Five Kingdoms,* 3rd edition. (See the Bibliography.)

The Ant's Journey

The levels of classification used in this chapter and throughout the book have been deliberately simplified. In actual taxonomic practice, there are often intermediate taxa between any two levels. Depending on the kind of organism being classified, there are subphyla and superclasses, subclasses and superorders, suborders and superfamiles, and so on. Even within species there are sometimes subspecies, varieties, and races.

Congress of Birds

The classic Persian work *Parliament of Birds,* by Fariduddin Attar, is echoed in a new context wherein even the basic elements of spirituality are absent. The great horned owl plays the role of the Hoopoe, a colorful, crested bird that was once common between the Tigris and the Hindu Kush. In Attar's work, the Hoopoe instructed the other birds in the higher stages of spiritual development. In the congress described here, the loftiest goal is survival, not spirituality. Yet the shadow of implicit blame falls on humanity for its inhumanity, a problem rooted in its lack of real spiritual connection.

Microperson

More than one popular exposition of microscopic life stresses the viscosity of water at micro-dimensions: a "life in syrup." In fact, the enormous energy available to miniature muscles, relatively speaking, more than compensates for the viscosity. Watch how a single whiplash flagellum, powered by a motor that is barely larger than the handle of the whip, drives a bacterium at enormous scale speeds.

The real problems and opportunities caused by shrinking are omitted from *Gulliver's Travels,* not to mention movies such as *Incredible Voyage.* Yet they make a wonderful platform from which to launch a real appreciation of the crucial differences between life-forms that occupy worlds on very different scales.

Water

The hydrogen bond not only gives water its cohesive properties but also helps complex protein molecules hold their shape. Stronger bonds are needed to form the molecules themselves, and these depend on shared pairs of electrons. In a molecule of carbonic acid, for example. the carbon atom shares a pair of electrons with each of two oxygens (the ones with the hydrogens attached). This is called a valence bond. With the remaining oxygen, the carbon shares two pairs of electrons, forming a double bond, which is not only stronger but also yields up more energy when it is broken.

The energy that all living things extract from their environment is stored in the valence bonds and double bonds of molecules within them.

Cymbella and the Hypotrich

It is to biologist Lynn Margulis that we owe the widespread understanding of our origins in the symbiosis that led to eucaryotic cells. By championing the symbiotic hypothesis, she enlarged our understanding of evolution as an interactive (rather than merely competitive) process.

The temporary symbiosis between *Cymbella* and the hypotrich is hypothesized by the author on the basis of hundreds of examples of living hypotrichs containing apparently healthy diatoms for long periods of time.

On the Back of a Turtle

The most primitive animals are thought to have evolved from colonial organisms like sponges, in which loosely coordinated and vaguely specialized cells cooperate to gather food, distribute it, and build a structure to house the colony. But the anatomy of rotifers and gastrotrichs (another aquatic animal of microscopic proportions) hints at another, quite different possibility. Perhaps the primitive animals were not multicellular but merely multinucleate. What better environment could there be for nuclei to learn the art of cooperation and coordination, after all, than one in which there were no walls?

The Meadow

Alternation of generations is one of the great principles of plant science. It is also one of the grand themes of evolution. An inconspicuous or barely useful organ, appendage, or strategy becomes, in time, an indispensable feature of the evolved organism. Pre-adaptation is the key to evolution.

A fuller picture of the gametophyte/sporophyte alternation would include more about the horsetails (Phylum Sphenophyta) as well as the conifers (Phylum Gymnospermatophyta).

Some biologists still describe the humble mosses and liverworts as though they were "evolutionary dead ends" or, at best, way stations on the road to something better. But mosses and liverworts are already perfectly suited to their smaller scale. The lack of a tracheal system may prevent them from becoming larger, but why should they do so? They are already the trees and bushes of World Four.

The Labyrinths

The standard analysis of soil recognizes four horizons: the litter layer, the A-layers (sand, silt, and clay permeated by organic material), the B-layer, (weathered parent rock with some leachate), and the C-layer (basement or country rock). The presence of a heavy clay layer does not fall easily into this scheme unless it is treated as the C-layer. Clays are more or less impervious to vertical leaching and are rarely the result of local weathering, having been transported miles from another site. In fact, most of the soil of the floodplain was brought there by a succession of spring floods over the last thousand years!

Lumbricus terrestris is not a native worm but an import from Europe. The native worms tend to be smaller, and they tunnel horizontally rather than vertically. Those in the northern half of the continent, including most of Canada and the northern states, were extirpated by glaciers. These lands were recolonized from the south and from Europe. The earthworms often found dead on sidewalks after rains did not drown but were killed by the ultraviolet light of day.

Prayer of the Mantis

No class of living things has so many species as the class Insecta. Fewer than a million species have been described by entomologists (including about 350,000 species of beetles alone) and estimates of the number of species yet to be described range as high as fifteen million! There are, at a guess, about a thousand entomologists out looking.

The Hydraulic Plant

The plant under stress illustrates the cohesion–tension theory of transpiration currently favored by plant physiologists. A previous theory, that capillarity alone accounts for the upward conduction of water in plant tissues, cannot explain the immense distances that water travels in trees and other tall plants.

The Storm

Meteorologists are still uncertain about the exact mechanism by which electric charges become separated inside storm clouds. An early theory holds that the descent of hail inside storm clouds acts as a giant Van de Graaff generator. A gradient of temperature from the center of a hailstone to its surface produces a slight gradient in charge as well, the outer surface of the stone being charged positively with respect to the center. As the hailstone falls, the outer rind melts away, shedding the positive charge and carrying the negative charge lower, into the bottom of the cloud. The multiplication of this slight effect by millions of hailstones accounts for the hundreds of thousands of volts that build up as potential.

Abundance

There is some uncertainty about the actual shape of the J-curve. The underlying mathematical formula (if there is one) is unknown, although many candidates have been proposed. Some of them have the basic shape of a hyperbola, the formula being $1/x$. It is fair to say that in most samples of natural communities (including tropical rainforests), the number of species of abundance x can be *estimated* as proportional to $1/x$, out to natural limits.

In the Forest

Earlier in the century, the native chestnuts all died as the result of a fungus imported from Europe. In the 1950s and 1960s, most of the elms died as the result of Dutch elm disease, another symbiotic insect/fungus attack. It is entirely possible that these species were also early, indirect victims of industrial pollution.

The Art of Decay

The classification of fungi used here is an uneasy compromise between the one proposed by microbiologist Lynn Margulis and that suggested by mycologist Bryce Kendrick. Margulis lists ascomycetes and basidiomycetes as separate phyla, whereas Kendrick makes them subphyla (Ascomycotina and Basidiomycotina) grouped into a single phylum called Dikaryomycota.

With Kendrick, this book recognizes only the zygomycetes, ascomycetes, and basidiomycetes as true fungi. As ultrastructural studies reveal new relationships among diverse organisms, classifications are being revamped, even at the top levels. For example, the protist phylum Chytridiomycota appears to be more closely related to true fungi than was previously thought, and it may be transferred from Kingdom Protista to Kingdom Fungi.

Actually, we do not know whether a raccoon would suffer the same consequences of eating a poisonous mushroom as a human would, assuming proportionate amounts. But I use Lotor (instead of Dianne) to illustrate the effects.

Bear

Few descendants of the European immigrants to this continent have ever conceived of the growing desperation and despair that gripped native North Americans as valley after valley and forest after forest were lost to the white onslaught. From the early sixteenth century to the end of the nineteenth, they fought a losing action against a people who honored no treaty, who came in mind-boggling numbers, and who treated the native North Americans just as they treated animals and trees, as something to be cleared away.

The tribe is made Shawnee in honor of Tecumseh, perhaps the greatest leader that the First People ever had. Tecumseh died not far from Hungry Hollow, in a manner of speaking, fighting the army of General (later President) Harrison.

Stories in Stone

The extent of glaciation varied greatly over what we now call the eastern deciduous forest. In the northern areas, glaciers developed a thickness of two kilometers. In the south there were no glaciers at all, but the flora and fauna, essentially boreal, were very different from today. In this chapter we strike a happy balance with a glacier that was only a few hundred meters thick.

Hungry Creek

If species of the lower worlds are incompletely known, their diets are almost completely unknown. Our understanding of ecology can never be better than our understanding of the underlying food webs, and our knowledge of the latter is at best partial. Laboratory experiments with specific animals or protists may sometimes tell us what these creatures may eat, at least under laboratory conditions, but who will extend their microscope by fiber-optic cable into the benthos to study the food preferences of the ultra-small?

Blooms of *Microcystis* are more common in ponds and lakes, but might occur in a streamfed stillwater environment.

Requiem for a Toad

Like all organisms that reproduce in great numbers to ensure their survival, toad populations must also die in great numbers. Gains and losses do not always balance, so populations fluctuate unpredictably. It can only be said that if food is plentiful and toads are not, then their numbers will sooner or later increase to match the supply, which by this time may have decreased again. Here lies the key to population dynamics, but it is so subject to the influence of other factors that mathematical models rarely fit well.

Survival of the Lucky

We do not know to what extent the DNA of a single-celled organism controls its behavior directly, but it is possible that instructions for what to do in a variety of circumstances are encoded in strings of messenger RNA that float within the cell.

The view of evolution presented here is somewhat non-standard in its emphasis on pre-adaptation. Paleontologists are increasingly persuaded, however, that many—if not all—speciation events began with smallish, isolated populations.

The Book of Kaolinite

I present here but one version of the theory of genetic takeover, as developed by Graham Cairns-Smith. In doing so, I do not claim it to be the only theory or even the best one. But the detailed speculation that it makes possible leads to more believable scenarios than the widely accepted speculation that life involved DNA or RNA from the very beginning and that it was synthesized and preserved in the early oceans by a sequence of miracles that stretch credulity ever thinner the more closely they are examined.

Didelphis virginianus

Didelphis virginianus herself was not around during the Triassic, or even the Jurassic. She is a modern animal, but one with very "early" features. A time traveler to the Jurassic could scarcely be blamed, on seeing her most direct ancestor, for thinking it a kind of possum.

Animal Minds

This chapter is almost entirely conjectural regarding Lotor's inner life, but the events of his outer life are typical.

Permanent Clearcut

No notes.

Ursa Major

Because this book addresses parallel situations in two countries, the politics of native affairs is made generic. Thus the names of legislation and of the band in question are fictional.

Bibliography

The books listed here touch upon some of the many themes presented in *Hungry Hollow*. There are focused readings on biology in general (including a great coloring book) and on specific areas of biology. There are books about various groups of organisms, as well as the most useful field guides (in the author's opinion) for them. All field guides apply either to eastern North America or to North America as a whole (sometimes excluding Mexico). There are books about the physical environment, from stones to storms. To use some of the field guides, it will be necessary to have a high-quality student microscope, which you can buy for half the price of a new computer—or less!

Ahmadjian, V., and Paracer, S. *Symbiosis: An Introduction to Biological Associations*. Hanover, NH: University Press of New England, 1986.

Atkins, P. W. *Molecules*. New York: Scientific American Library, 1987.

Borror, D. J., and White, R. E. A *Field Guide to Insects*. The Peterson Field Guide Series. Boston: Houghton Mifflin, 1970.

Burgess, J.; Marten, M.; and Taylor, R. *Under the Microscope: A Hidden World Revealed*. New York: Cambridge University Press, 1990.

Buschbaum, R. *Animals Without Backbones*. 3d ed. Chicago: The University of Chicago Press, 1987.

Cairns-Smith, A. G. *Genetic Takeover and the Mineral Origins of Life.* New York: Cambridge University Press, 1982.

Cobb, B. *A Field Guide to the Ferns.* The Peterson Field Guide Series. Boston: Houghton Mifflin, 1984.

Conant, R., and Collins, J. T. *A Field Guide to Reptiles and Amphibians: Eastern and Central North America.* 3d ed. The Peterson Field Guide Series. Boston: Houghton Mifflin, 1991.

Conard, H. S., and Redfern, P. L., Jr. *How to Know the Mosses and Liverworts.* 2d ed. The Pictured Key Nature Series. Dubuque, IA: Wm. C. Brown, 1979.

Cotton, W. R. *Storms.* Fort Collins, CO: Aster Press, 1990.

Dowden, A. O. *From Flower to Fruit.* New York: Ticknor and Fields, 1984.

Emiliani, C. *The Scientific Companion: Exploring the Physical World with Facts, Figures, and Formulas.* New York: Wiley, 1988.

Flint, R. F., and Skinner, B. J. *Physical Geology.* New York: Wiley, 1974.

Gould, J. L.; Keeton, W. T.; and Gould, C. G. *Biological Science,* vol. 1 and vol. 2. 6th ed. New York: Norton, 1996.

Griffin, D. R. *Animal Minds.* Chicago: University of Chicago Press, 1992.

Griffin, R. D. *The Biology Coloring Book.* New York: Harper-Collins, 1986.

Gullan, P. J., and Cranston, P. S. *The Insects: An Outline of Entomology.* Chapman and Hall, London, 1994.

Heywood, V. H., exec. ed. *Global Biodiversity Assessment.* London: United Nations Environment Programme/Cambridge University Press, 1995.

Hopkins, W. G. *Introduction to Plant Physiology.* New York: Wiley, 1995.

Horst, R. K. *Westcott's Plant Disease Handbook*. 5th ed. New York: Van Nostrand Reinhold, 1990.

Jahn, T. L.; Bovee, E. C.; and Jahn, F. F. *How to Know the Protozoa* 2d ed. The Pictured Key Nature Series. Dubuque, IA: Wm. C. Brown, 1979.

Kendrick, B. *The Fifth Kingdom*. 2d ed. Newburyport, MA: Focus Texts, 1992.

Lincoff, G. H. *The Audubon Society Field Guide to North American Mushrooms*. New York: Knopf, 1991.

Margulis, L., and Sagan, D. *Microcosmos: Four Billions Years of Evolution from Our Microbial Ancestors*. New York: Simon and Schuster, 1986.

Margulis, L., and Schwartz, K. V. *Five Kingdoms: An Illustrated Guide to the Phyla of Life on Earth*. 3rd ed. New York: Freeman, 1998.

Milne, L., and Milne, M. *The Audubon Society Field Guide to the North American Insects and Spiders*. New York: Knopf, 1984.

Newcombe, L. *Newcombe's Windflower Guide*. Boston: Little, Brown, 1977.

Niering, W. A., and Olmstead, N. C. *National Audubon Society Field Guide to North American Wildflowers*. New York: Knopf, 1997.

Page, L. M., and Burr, B. M. *A Field Guide to Freshwater Fishes*. The Peterson Field Guide Series. Boston: Houghton Mifflin, 1991.

Petrides, G. A. *A Field Guide to the Eastern Trees*. The Peterson Field Guide Series. Boston: Houghton Mifflin, 1988.

Prescott, G. W. *How to Know the Freshwater Algae*. 3d ed. The Pictured Key Nature Series. Dubuque, IA: Wm. C. Brown, 1978.

Robbins, C. S.; Brunn, B.; and Zim, H. S. *Birds of North America*. New York: Golden Press, 1983.

Thorp, J. H., and Covich, A. P., eds. *Ecology and Classification of North American Freshwater Invertebrates.* San Diego, CA: Academic Press, 1991.

Tuttle, M. D. *America's Neighborhood Bats.* Austin, TX: University of Texas Press, 1988.

Wood, M. *Soil Biology* New York: Chapman and Hall, 1989.